Fluid and
Solid Mechanics

LTCC Advanced Mathematics Series

Series Editors: Shaun Bullett *(Queen Mary University of London, UK)*
Tom Fearn *(University College London, UK)*
Frank Smith *(University College London, UK)*

LTCC Advanced Mathematics Series - Volume 2

Fluid and Solid Mechanics

Editors

Shaun Bullett
Queen Mary University of London, UK

Tom Fearn
University College London, UK

Frank Smith
University College London, UK

NEW JERSEY · LONDON · SINGAPORE · BEIJING · SHANGHAI · HONG KONG · TAIPEI · CHENNAI · TOKYO

Published by

World Scientific Publishing Europe Ltd.

57 Shelton Street, Covent Garden, London WC2H 9HE

Head office: 5 Toh Tuck Link, Singapore 596224

USA office: 27 Warren Street, Suite 401-402, Hackensack, NJ 07601

Library of Congress Cataloging-in-Publication Data
Names: Bullett, Shaun, 1967– | Fearn, T., 1949– | Smith, F. T. (Frank T.), 1948–
Title: Fluid and solid mechanics / [edited by] Shaun Bullett (Queen Mary
  University of London, UK), Tom Fearn (University College London, UK) &
  Frank Smith (University College London, UK).
Description: New Jersey : World Scientific, 2016. | Series: LTCC advanced mathematics series ;
  volume 2 | Includes bibliographical references and index.
Identifiers: LCCN 2015049962| ISBN 9781786340252 (hc : alk. paper) |
  ISBN 9781786340269 (pbk : alk. paper)
Subjects: LCSH: Mechanics, Analytic. | Fluid mechanics.
Classification: LCC QA805 .F54 2016 | DDC 531--dc23
LC record available at http://lccn.loc.gov/2015049962

British Library Cataloguing-in-Publication Data
A catalogue record for this book is available from the British Library.

Desk Editors: R. Raghavarshini/Mary Simpson

Typeset by Stallion Press
Email: enquiries@stallionpress.com

Printed in Singapore

Preface

The London Taught Course Centre (LTCC) for PhD students in the Mathematical Sciences has the objective of introducing research students to a broad range of topics. For some students, some of these topics might be of obvious relevance to their PhD projects, but the relevance of most will be much less obvious or apparently non-existent. However, all of us involved in mathematical research have experienced that extraordinary moment when the penny drops and some tiny gem of information from outside ones immediate research field turns out to be the key to unravelling a seemingly insoluble problem, or to opening up a new vista of mathematical structure. By offering our students advanced introductions to a range of different areas of mathematics, we hope to open their eyes to new possibilities that they might not otherwise encounter.

Each volume in this series consists of chapters on a group of related themes, based on modules taught at the LTCC by their authors. These modules were already short (five two-hour lectures) and in most cases the lecture notes here are even shorter, covering perhaps three-quarters of the content of the original LTCC course. This brevity was quite deliberate on the part of the editors — we asked the authors to confine themselves to around 35 pages in each chapter, in order to allow as many topics as possible to be included in each volume, while keeping the volumes digestible. The chapters are "advanced introductions", and readers who wish to learn more are encouraged to continue elsewhere. There has been no attempt to make the coverage of topics comprehensive. That would be impossible in any case — any book or series of books which included all that a PhD student in mathematics might need to know would be so large as to be totally unreadable. Instead what we present in this series is a cross-section of some of the topics, both classical and new, that have appeared in LTCC modules in the nine years since it was founded.

The present volume is within the area of fluid and solid mechanics. The main readers are likely to be graduate students and more experienced researchers in the mathematical sciences, looking for introductions to areas with which they are unfamiliar. The mathematics presented is intended to be accessible to first year PhD students, whatever their specialised areas of research. Whatever your mathematical background, we encourage you to dive in, and we hope that you will enjoy the experience of widening your mathematical knowledge by reading these concise introductory accounts written by experts at the forefront of current research.

Shaun Bullett, Tom Fearn, Frank Smith

Contents

Chapter 1

Introductory Geophysical Fluid Dynamics

Michael Davey

Department of Applied Mathematics and Theoretical Physics,
University of Cambridge, Centre for Mathematical Sciences,
Wilberforce Road, Cambridge CB3 0WA, UK
mkd3@cam.ac.uk

This chapter concerns mathematical modelling of large-scale fluid flows relative to a rotating frame of reference, for which the effects of rotation are dominant and to leading order there is a balance of horizontal pressure gradients and Coriolis forces. The principal application is to oceanic and atmospheric flows with horizontal scales of tens of kilometres or more, and timescales of days or more. A fundamental equation in the dynamics of such flows is that for quasigeostrophic potential vorticity, and this is derived in the first part of the chapter, with stratification effects included in the form of layers with constant density within each layer. Large-scale wave-like behaviour is supported in the form of Rossby waves, and some basic properties of these waves are presented. Simplified conceptual quasigeostrophic models provide understanding of dynamical processes, and two examples are described: ocean spin-up and multiple equilibria.

1. Introduction

Mathematical representation of large-scale atmospheric and oceanic flows has great practical importance as it provides the basis for the dynamical numerical models used for making weather and climate outlooks for hours to decades ahead. The full equations of fluid motion are too complex to use for this purpose, but mathematical theory provides the foundation for approximations that represent the scales and phenomena of interest and allow efficient numerical computation. Even with these approximations, atmospheric and oceanic flows contain processes and interactions on a wide range of space and time scales. Mathematical models can further be used to focus on particular processes and investigate their behaviour and roles.

This chapter contains a subset of a course intended for graduates who are familar with the basics of fluid mechanics, such as the Navier–Stokes equations and wave-like behaviour such as gravity waves, but have not encountered geophysical fluid dynamics. A brief explanation of the governing equations for quasigeostrophic flow is provided, without rigorous justification for the various standard approximations employed.

Two examples are provided of conceptual models based on the quasigeostrophic potential vorticity equations. One is a model of mid-latitude wind-driven ocean circulation. The classic steady case demonstrates how intense currents such as the Gulf Stream occur near western boundaries, while the time-dependent part illustrates how Rossby waves influence ocean circulation and how in a stratified ocean they provide oceans with a long-term "memory". The second example demonstrates how the interaction of Rossby waves, topographic drag and mean flow may create multiple stable states, relevant in particular to "blocked" flow regimes in the atmosphere.

There are many good textbooks on this subject. More detailed and rigorous derivations of sets of equations relevant to geophysical fluid dynamics, with applications, may be found in books by Gill,[1] Pedlosky[2] and Vallis[3] for example.

2. Governing Equations

For flows relative to a rotating frame of reference, the Navier–Stokes equations have the form

$$\frac{\mathrm{D}\underline{u}}{\mathrm{D}t} + 2\underline{\Omega} \times \underline{u} = -\frac{\nabla p}{\rho} + \nu \nabla^2 \underline{u} + \text{gravitational effects}, \qquad (1)$$

where $\underline{u}$ is the velocity vector, p is pressure, ρ is density, $\mathrm{D}/\mathrm{D}t$ indicates a derivative following the motion and ν is a viscosity coefficient.

For planet Earth, the rotation vector $\underline{\Omega}$ has magnitude $\Omega = 2\pi$ radians per day and direction outward from the North Pole. Earth can be regarded as a sphere of radius R_e, with the atmospheric and oceanic flows in thin layers near that radius, with large horizontal scale compared to the depth in each medium.

With flows in mid-latitude regions in mind, choose a coordinate system that is centred on some latitude θ_0. For simplicity, locally Cartesian coordinates are defined, with x in the zonal (west to east) direction, y in the meridional (south to north) direction, and z vertically upwards. The corresponding fluid velocity components are denoted u (zonal), v (meridional) and w (upward). Note that $y = R_e(\theta - \theta_0)$, and at latitude θ the radially outward component of the rotation vector has size $\Omega \sin \theta$.

The horizontal momentum equations are

$$u_t + (\underline{u} \cdot \nabla)u - fv = -\frac{p_x}{\rho} + \nu \nabla^2 u, \qquad (2a)$$

$$v_t + (\underline{u} \cdot \nabla)v + fu = -\frac{p_y}{\rho} + \nu \nabla^2 v. \qquad (2b)$$

Here, $f = 2\Omega \sin \theta$ is referred to as the Coriolis parameter. Other components of the rotation vector other than f have small influence, and have been omitted in (2). (For brevity formal justifications of the approximations made here and elsewhere are omitted: details can be found in textbooks such as those cited in the introduction.)

The variation of f with latitude is fundamental to many properties of the large-scale flows of interest here. A further very common simplification is to consider f as a linear function of y, by using $\sin \theta \approx \sin \theta_0 + (\theta - \theta_0) \cos \theta_0$. Then $f \approx f_0 + \beta y$, where $f_0 = 2\Omega \sin \theta_0$ and $\beta = 2\Omega \cos \theta_0 / R_e$. With this assumption, the coordinate system is known as a "beta-plane". In standard terminology, a system with $\beta = 0$ is referred to as an "f-plane".

Mass conservation requires

$$\rho_t + \nabla \cdot (\rho \underline{u}) = 0. \qquad (3)$$

2.1. *Hydrostatic balance*

For oceanic and atmospheric flows with horizontal scale much larger than the vertical scale, the vertical equation of motion is dominated by the balance of vertical pressure gradient and gravitational force. To a very good approximation, the system is in hydrostatic balance, with

$$p_z = -\rho g. \tag{4}$$

Effectively the pressure at any point is determined by the mass of overlying fluid, and is not influenced by the fluid motion.

2.2. *Geostrophic balance*

Suppose the flow has length scale L, a horizontal velocity scale U and an advective time scale L/U. The non-dimensional Rossby number R, fundamental to rotating flows, is defined as

$$R = \frac{U}{(f_0 L)}. \tag{5}$$

We assume the Rossby number is small ($R \ll 1$), in which case the left-hand side of (2) is dominated by the Coriolis terms fv and fu. Apart from thin layers near boundaries, viscous and forcing effects are small. The dominant balance is between the Coriolis and pressure gradient terms:

$$-fv = -\frac{p_x}{\rho}, \quad fu = -\frac{p_y}{\rho}. \tag{6}$$

This is referred to as "geostrophic balance". For later use, define a geostrophic horizontal flow u_g, v_g by

$$-v_g = -\frac{p_x}{(\rho_0 f_0)}, \quad u_g = -\frac{p_y}{(\rho_0 f_0)}, \tag{7}$$

where ρ_0 is a typical density scale. Note that $\underline{u}_g \cdot \nabla p = 0$: the geostrophic flow follows lines of constant pressure, i.e., isobars. Flow is cyclonic around low pressure centres, and anti-cyclonic around high pressure centres. In the northern hemisphere, orientation is such that cyclonic flow is anti-clockwise. Note also that $u_{gx} + v_{gy} = 0$, so the geostrophic flow is horizontally non-divergent. Thus a geostrophic streamfunction ψ can be defined: conventionally such that

$$u_g = -\psi_y, \quad v_g = \psi_x. \tag{8}$$

(Note: from here on assume ∇ is the horizontal gradient operator, and $\underline{u} = (u, v)$, as should be obvious from the context.)

2.3. *Representation of the density distribution as layers*

In practice, density varies throughout the ocean and atmosphere. The ideal gas law applies to the atmosphere, and in the ocean density depends mainly on temperature and salinity. As is often done for conceptual models and theoretical investigations, we will consider a simple representation of the density structure as layers in which each layer has a constant prescribed density. (See textbooks for alternative representations.) Thus, thermodynamic aspects that influence the density are not considered here: instead, the focus is on dynamic processes. We shall also assume that density varies little throughout the system. This is a viewpoint more directly suited to the oceans than the atmosphere: for the latter an alternative formulation can be made that allows for the substantial vertical variation of air density through the depth of the atmosphere at rest.

Suppose a system has N layers, with layer 1 at the top overlying layer 2 overlying layer 3, etc. Suppose layer n has density ρ_n, with $\rho_n < \rho_{n+1}$. With the fluid at rest the interfaces are horizontal, and the layers have depths H_n which are constant, except for the lowest layer whose depth may vary with x and y to allow the possibility of underlying topography. Suppose the top of layer 1 is at $z = z_T$ when undisturbed, and suppose the perturbation of this surface is η_1, so the top of disturbed layer 1 is at $z_T + \eta_1$. Similarly, η_2 denotes the perturbation to the interface between layers 1 and 2, so the bottom of layer 1 (and the top of layer 2) is at $z_T - H_1 + \eta_2$, and so on. A fixed perturbation $\eta_{N+1}(x, y)$ at the base of layer N can be used to represent the topography, so the bottom of layer N is at $z_T - (H_1 + \cdots + H_N) + \eta_{N+1}$. The density structure for such a system with two layers is illustrated in Fig. 1.

Using hydrostatic balance, some useful relationships between pressures and layer distributions can be derived. Suppose the pressure at the top surface is p_T at $z = z_T$. From the hydrostatic relation, in layer 1

$$p_1 = p_T + \rho_1 g \eta_1 - \rho_1 g(z - z_T) \tag{9}$$

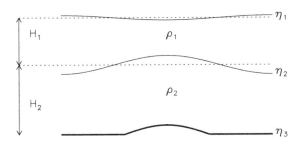

Fig. 1.   Schematic diagram of the density structure in a two-layer system.

for $z_T - H_1 + \eta_2 < z < z_T + \eta_1$, and in layer 2

$$p_2 = p_T + \rho_1 g \eta_1 + (\rho_1 - \rho_2) g (H_1 - \eta_2) - \rho_2 g (z - z_T) \qquad (10)$$

for $z_T - H_1 - H_2 + \eta_3 < z < z_T - H_1 + \eta_2$, etc.

Suppose p_T is constant. (For oceanic applications, p_T is the sea level atmospheric pressure, fluctuations in which are small compared to pressure fluctuations below the surface and negligible for most circumstances. For atmospheric applications, p_T is effectively zero at the top of the atmosphere.) Then the horizontal pressure gradients are independent of depth within each layer:

$$\nabla p_1 = \rho_1 g \nabla \eta_1, \quad \nabla p_2 = \rho_1 g \nabla \eta_1 + (\rho_2 - \rho_1) g \nabla \eta_2, \text{ etc.} \qquad (11)$$

Thus, the geostrophic flow $\underline{u}_{g1}$ in layer 1, determined by ∇p_1, can be diagnosed from η_1 and is independent of depth in layer 1; likewise $\underline{u}_{g2}$ is determined by η_1 and η_2, etc. (Thus $\underline{u}_g$ is determined by the overlying density structure.) Note that

$$\nabla (p_2 - p_1) = (\rho_2 - \rho_1) g \nabla \eta_2, \qquad (12)$$

so the difference $\underline{u}_{g1} - \underline{u}_{g2}$ is determined by η_2. The vertical geostrophic shear between two layers is determined by the horizontal gradient in the intervening density structure.

With density constant within each layer, from (3) it follows that

$$u_x + v_y + w_z = 0. \qquad (13)$$

2.4. *Shallow water equations and potential vorticity*

The above relations show how the density structure, pressure gradients and geostrophic flows are diagnostically related. (In particular, given all the ρ_n and η_n, $\underline{u}_g$ are known.) However, more information is needed to find out how they evolve. The relevant equations can be derived rigorously through asymptotic expansions with the Rossby number as a small parameter. Here, a more *ad hoc* approach is adopted, making a series of assumptions that can ultimately be justified more formally.

Within each layer (away from possible thin frictional layers at the boundaries) the horizontal flow is independent of depth and governed by the "shallow water equations":

$$u_t + (\underline{u} \cdot \nabla)u - fv = -\frac{p_x}{\rho_0} + A\nabla^2 u, \tag{14a}$$

$$v_t + (\underline{u} \cdot \nabla)v + fu = -\frac{p_y}{\rho_0} + A\nabla^2 v. \tag{14b}$$

The situation is analogous to long waves (wavelength much greater than depth) in shallow water, hence the name. Note that a constant reference density is used in the pressure gradient terms, valid for all layers. Here, A is a weak horizontal diffusivity coefficient, whose influence is negligible except in regions of strong gradients. (This diffusivity is intended to represent the effects of small scales for which the approximate equations are no longer valid, rather than molecular viscosity.)

The vorticity (more correctly, the vertical component of the vorticity vector) is $\zeta = v_x - u_y$. The vector identity $(\underline{u} \cdot \nabla)\underline{u} = \nabla \underline{u}^2/2 - \underline{u} \times \zeta\underline{k}$, where $\underline{k}$ is the unit vertical vector, can be used to write (14) as

$$u_t + \frac{1}{2}(\underline{u}^2)_x - (f + \zeta)v = -\frac{p_x}{\rho_0} + A\nabla^2 u, \tag{15a}$$

$$v_t + \frac{1}{2}(\underline{u}^2)_y + (f + \zeta)u = -\frac{p_y}{\rho_0} + A\nabla^2 v. \tag{15b}$$

Eliminating p and $\underline{u}^2$ by cross-differentiating leads to the vorticity equation

$$\zeta_t + \underline{u} \cdot \nabla(\zeta + f) + (\zeta + f)\nabla \cdot \underline{u} = A\nabla^2\zeta. \tag{16}$$

From (13), it follows that

$$\frac{D(\zeta + f)}{Dt} = (\zeta + f)w_z + A\nabla^2\zeta, \tag{17}$$

where the material derivative is $D/Dt = \partial/\partial t + \underline{u} \cdot \nabla$. Thus changes in the total vorticity $\zeta + f$ are induced by vertical motion stretching or shrinking the fluid column within the layer and by dissipation.

As u and v are independent of depth within the layer, it follows from (13) that w_z is depth-independent and hence

$$w_z = \frac{(w_T - w_B)}{h}, \tag{18}$$

where h denotes the layer depth and w_T and w_B denote w near the top and bottom of the layer.

The vertical motions w_T and w_B are influenced both by the upper and lower layer boundary positions and by thin frictional layers known as Ekman layers. Within these thin boundary layers, the flow is adjusted to match boundary conditions at the layer boundaries where necessary. For brevity, the standard properties of Ekman layers are not derived here, but will simply be stated as required in later sections. For now, write

$$w_T - w_B = \frac{Dh}{Dt} + w_{ET} - w_{EB}, \tag{19}$$

indicating the contributions from layer boundary positions and from the thin Ekman layers.

Noting that

$$\frac{D}{Dt}\frac{(\zeta + f)}{h} = \frac{1}{h}\frac{D}{Dt}(\zeta + f) - \frac{(\zeta + f)}{h^2}\frac{D}{Dt}h,$$

it follows from (17) and (18) that

$$\frac{D}{Dt}\frac{(\zeta + f)}{h} = \frac{(\zeta + f)}{h}\frac{(w_{ET} - w_{EB})}{h} + \frac{1}{h}A\nabla^2\zeta. \tag{20}$$

The expression $(\zeta + f)/h$ is the potential vorticity, which is a fundamental quantity in geophysical fluid dynamics. (There are equivalent expressions in continuously stratified systems.)

2.5. *Quasigeostrophic potential vorticity*

Suppose the layer displacements are small compared to the layer depths, so

$$\frac{1}{h_n} = \frac{1}{(H_n + \eta_n - \eta_{n+1})} \approx \left(\frac{1}{H_n}\right)\left[1 - \frac{(\eta_n - \eta_{n+1})}{H_n}\right]. \qquad (21)$$

With $\zeta_n + f = \zeta_n + f_0 + \beta y$, the potential vorticity is approximately

$$\frac{(\zeta_n + f)}{h_n} \approx \left(\frac{f_0}{H_n}\right)\left[1 + \frac{(\zeta_n + \beta y)}{f_0}\right]\left[1 - \frac{(\eta_n - \eta_{n+1})}{H_n}\right].$$

Assuming $\zeta_n + \beta y$ is small compared to f_0, and neglecting the product of the small terms, we have

$$\frac{(\zeta_n + f)}{h_n} \approx \left(\frac{f_0}{H_n}\right)\left[1 + \frac{(\zeta_n + \beta y)}{f_0} - \frac{(\eta_n - \eta_{n+1})}{H_n}\right]. \qquad (22)$$

Further, the flow in the layer is approximately geostrophic: $\underline{u} \approx \underline{u}_g$. Thus, from (20), we obtain to leading order the quasi-geostrophic potential vorticity equations for a layered system:

$$\frac{D_g q_n}{Dt} = \left(\frac{f_0}{H_n}\right)(w_{EnT} - w_{EnB}) + A\nabla^2 \zeta_{gn}, \qquad (23)$$

where

$$q_n = \zeta_{gn} + \beta y - \left(\frac{f_0}{H_n}\right)(\eta_n - \eta_{n+1}) \qquad (24)$$

with geostrophic vorticity $\zeta_{gn} = v_{gnx} - u_{gny}$ and $D_g/Dt = \partial/\partial t + \underline{u}_g \cdot \nabla$. Note that in terms of the streamfunction introduced by (8), $\zeta_{gn} = \nabla^2 \psi_n$.

The streamfunctions and interface displacements are related: e.g., using (7) and (11), etc.

$$\psi_1 = \frac{g\eta_1}{f_0}, \quad \psi_n = \psi_{n-1} + \frac{g'_n \eta_n}{f_0}, \text{ etc.,} \qquad (25)$$

where $g'_n = g(\rho_n - \rho_{n-1})/\rho_0$ is called "reduced gravity".

Thus, the quasigeostrophic potential vorticities q_n can be written as:

$$q_1 = \nabla^2 \psi_1 + \beta y - \frac{f_0^2}{H_1} \left[\frac{\psi_1}{g} - \frac{(\psi_2 - \psi_1)}{g_2'} \right], \tag{26a}$$

$$q_n = \nabla^2 \psi_n + \beta y - \frac{f_0^2}{H_n} \left[\frac{\psi_n - \psi_{n-1}}{g_n'} - \frac{(\psi_{n+1} - \psi_n)}{g_{n+1}'} \right], \tag{26b}$$

$$q_N = \nabla^2 \psi_N + \beta y - \frac{f_0^2}{H_N} \left[\frac{\psi_N - \psi_{N-1}}{g_N'} \right] + \frac{f_0}{H_N} \eta_{N+1}. \tag{26c}$$

3. Vertical Modes and Rossby Waves in a Two-Layer System

Variations in the Coriolis parameter f with latitude provide an important mechanism for large-scale waves in the ocean and atmosphere. We illustrate this by considering a two-layer system on a β-plane, ignoring the effects of dissipation, forcing and topography to focus on the Rossby waves. Further, linearise the quasigeostrophic potential vorticity equations by omitting terms of the form $\underline{u} \cdot \nabla \zeta$ and $\underline{u} \cdot \nabla \eta$ which involve multiples of derivatives of ψ, to obtain from (23) and (26)

$$\frac{\partial}{\partial t} \left[\nabla^2 \psi_1 - \frac{f_0^2}{g H_1} \psi_1 + \frac{f_0^2}{g' H_1} (\psi_2 - \psi_1) \right] + \beta \psi_{1x} = 0, \tag{27a}$$

$$\frac{\partial}{\partial t} \left[\nabla^2 \psi_2 - - \frac{f_0^2}{g' H_2} (\psi_2 - \psi_1) \right] + \beta \psi_{2x} = 0. \tag{27b}$$

Here, $g' = g \Delta \rho / \rho_0$ with $\Delta \rho = \rho_2 - \rho_1$. It is convenient to look for the vertical "modes" which have coherent behaviour in each layer. While not strictly necessary, some further assumptions are made to simplify the algebra.

3.1. *Baroclinic mode*

The internal density structure allows the geostrophic flows to vary with depth in the baroclinic mode. (In a system with several layers there may be several baroclinic modes: in this two-layer example there is just one.) Taking the difference of the equations (27) gives

$$\nabla^2 \hat{\psi}_t - \frac{f_0^2}{g'} \left(\frac{1}{H_1} + \frac{1}{H_2} \right) \hat{\psi}_t - \frac{f_0^2}{gH_1} \psi_{1t} + \beta \hat{\psi}_x = 0, \qquad (28)$$

where $\hat{\psi} = \psi_1 - \psi_2$ is a streamfunction for the baroclinic mode, and for the geostrophic shear $\underline{u}_{g1} - \underline{u}_{g2}$. For this mode, the surface displacement η_1 is much less than that of the interface displacement η_2, so the term $(f_0^2/gH_1)\psi_{1t}$ can be omitted to obtain

$$\nabla^2 \hat{\psi}_t - \frac{\hat{\psi}_t}{a^2} + \beta \hat{\psi}_x = 0, \qquad (29)$$

where the term

$$a^2 = \frac{g' H_1 H_2}{f_0^2 (H_1 + H_2)}, \qquad (30)$$

defines a length scale a known *inter alia* as the "internal Rossby radius" or "internal deformation scale" or "baroclinic Rossby radius". This scale is the distance travelled by an internal gravity wave with speed $\sqrt{g' H_1 H_2/(H_1 + H_2)}$ in time $1/f_0$, which is a scale valid also for f-plane motion. For the baroclinic mode, internal density variations (manifest here by variations in the interface displacement) are essential.

3.2. *Barotropic mode*

Taking the depth-weighted sum of the Eqs. (27) leads instead to

$$\nabla^2 (H_1 \psi_1 + H_2 \psi_2)_t - \frac{f_0^2}{g} \psi_{1t} + \beta (H_1 \psi_1 + H_2 \psi_2)_x = 0. \qquad (31)$$

Define a barotropic streamfunction by

$$\bar{\psi} = \frac{(H_1 \psi_1 + H_2 \psi_2)}{(H_1 + H_2)} \qquad (32)$$

which is the streamfunction for the depth-averaged geostrophic flow. For the barotropic mode, the geostrophic flow is the same in both layers to a good approximation. Putting $\psi_1 = \psi_2 = \bar{\psi}$, (31) can be written as

$$\nabla^2 \bar{\psi}_t - \frac{\bar{\psi}_t}{\bar{a}^2} + \beta \bar{\psi}_x = 0, \tag{33}$$

where

$$\bar{a}^2 = \frac{g(H_1 + H_2)}{f_0^2}, \tag{34}$$

defines a scale $\bar{a}$ analagous to a, but with reference to the "external" gravity wave speed $\sqrt{g(H_1 + H_2)}$. The barotropic mode behaves as though the fluid were not stratified. The barotropic adjustment scale is larger than the baroclinic scale a, and in practice the term $\bar{\psi}_t/\bar{a}^2$ can often be neglected in (33).

3.3. *Rossby waves*

For the baroclinic equation (29), consider a solution of the form

$$\hat{\psi} = A e^{i(kx+ly-\omega t)}, \tag{35}$$

where A is an arbitrary amplitude, k is zonal wavenumber, l is meridional wavenumber and ω is frequency. Then $\nabla^2 \hat{\psi} = -K^2 \hat{\psi}$, where $K^2 = k^2 + l^2$, and $\hat{\psi}_x = -ik\hat{\psi}$, etc. so (29) leads to the dispersion relation

$$\omega = -\frac{\beta a^2 k}{(K^2 a^2 + 1)}. \tag{36}$$

Thus, there are wave-like solutions, known as Rossby waves, when β is non-zero. The zonal phase speed of these baroclinic Rossby waves is

$$\frac{\omega}{k} = -\frac{\beta a^2}{(K^2 a^2 + 1)} \tag{37}$$

which is always negative, i.e., westward. Zonal propagation of information by these waves is determined by the zonal group velocity, which is

$$\frac{\partial \omega}{\partial k} = \frac{\beta a^2 [(k^2 - l^2)a^2 - 1]}{(K^2 a^2 + 1)^2}. \tag{38}$$

This may be positive or negative: waves with $k^2 a^2 < l^2 a^2 + 1$ have negative (westward) zonal group velocity while longer waves with

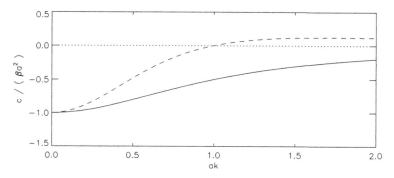

Fig. 2. Zonal phase velocity (solid) and group velocity (dashed) for Rossby waves with meridional wavenumber $l = 0$, as a function of zonal wavenumber k scaled by Rossby radius a.

$k^2 a^2 > l^2 a^2 + 1$ have eastward zonal group velocity. The dependence of phase and group velocity on k is illustrated in Fig. 2 for the case $al = 0$.

Note that for long waves (ka and $la \ll 1$; wavelength large compared to a)

$$\frac{\omega}{k} \approx -\beta a^2, \quad \frac{\partial \omega}{\partial k} \approx -\beta a^2. \tag{39}$$

Thus, long Rossby waves are non-dispersive and propagate westward: these are also the fastest Rossby waves.

Equation (33) has the same form as (29), and has analogous barotropic Rossby wave solutions: just replace a by $\bar{a}$ in the above results for dispersion and wave speed.

If the surface displacement is ignored (the so-called "rigid lid" approximation; effectively $\bar{a} \to \infty$), then (33) has Rossby wave solutions with

$$\frac{\omega}{k} = -\frac{\beta}{K^2}, \quad \frac{\partial \omega}{\partial k} = \frac{\beta(k^2 - l^2)}{K^4}. \tag{40}$$

3.4. *Oceanic scales*

The contrast between baroclinic and barotropic scales and speeds is particularly marked for the ocean. Representative values for a mid-latitude ocean (around say 40°N) are: $H_1 = 1000\,\text{m}$, $H_2 = 4000\,\text{m}$,

$f_0 = 9.4 \times 10^{-5}\,\mathrm{s}^{-1}$, $\beta = 1.75 \times 10^{-11}\mathrm{m}^{-1}\mathrm{s}^{-1}$, $\rho = 1000\,\mathrm{kgm}^{-3}$, $\Delta\rho/\rho = 0.003$. For the baroclinic mode, the Rossby radius a is then about $50\,\mathrm{km}$, and the fastest long Rossby wave speed βa^2 is about $5\,\mathrm{cm/s}$, while for the barotropic mode the Rossby radius is about $2300\,\mathrm{km}$ and the fastest long Rossby wave speed is about $100\,\mathrm{m/s}$.

4. Wind-Driven Mid-Latitude Ocean Circulation

To illustrate some basic properties of mid-latitude wind-driven strati-fied ocean circulation, consider a highly idealised "textbook" scenario using a two-layer β-plane system. The model ocean has a western boundary at $x_W = 0$ and eastern boundary at $x_E = L$, extending meridionally from $y = -L/2$ to $y = L/2$. The system is forced by surface wind stress, and damped by lateral dissipation A. The wind-driven mid-latitude ocean model with lateral dissipation was first introduced by Munk.[4]

4.1. *Governing equations and boundary conditions*

The wind stress drives the underlying ocean via the thin frictional surface Ekman layer. While the wind directly drives horizontal cur-rents in this surface layer, it also drives the underlying ocean by generating pressure gradients (through small but large-scale changes in sea level) that influence the deeper ocean and by driving vorticity changes via the vertical velocity w_{E1T}. In particular, we state with-out proof that within layer 1 the vertical velocity w_{E1T} is related to the wind stress curl by

$$w_{E1T} = \frac{(\tau_x^{(y)} - \tau_y^{(x)})}{(\rho f_0)}, \tag{41}$$

where $\tau^{(x)}$ and $\tau^{(y)}$ are the zonal and meridional components of the wind stress. Other possible Ekman layer contributions are omitted in this model, so $w_{E1B} = w_{E2T} = w_{E2B} = 0$.

For simplicity, assume "rigid lid" dynamics and flat topography, and linearise the quasigeostrophic vorticity equations about a state of rest. The governing equations are (from (23), neglecting terms quadratic in ψ)

$$\nabla^2\psi_{1t} + \beta\psi_{1x} + a_1^{-2}(\psi_2 - \psi_1)_t = \frac{1}{\rho H_1}(\tau_x^{(y)} - \tau_y^{(x)}) + A\nabla^4\psi_1,$$
(42a)

$$\nabla^2\psi_{2t} + \beta\psi_{2x} - a_2^{-2}(\psi_2 - \psi_1)_t = A\nabla^4\psi_2,$$
(42b)

where

$$a_1^2 = \frac{g'H_1}{f_0^2}, \quad a_2^2 = \frac{g'H_2}{f_0^2}.$$
(43)

For a northern hemisphere mid-latitude ocean such as the North Atlantic the prevailing surface winds are westerly (toward the east) in the northern half and easterly in the southern half of the region: a simple representation of this is

$$\tau^{(x)} = \tau_0 \sin\left(\frac{\pi y}{L}\right), \quad \tau^{(y)} = 0$$
(44)

with $\tau_0 > 0$. Then the wind stress curl is

$$\tau_x^{(y)} - \tau_y^{(x)} = -\tau_0\left(\frac{\pi}{L}\right)\cos\left(\frac{\pi y}{L}\right).$$
(45)

Effectively, the wind stress imposes a downward motion w_{E1T} at the surface of the quasigeostrophic system, which is largest at the central latitude $y = 0$.

The streamfunctions in this linear model have the same dependence on y as the wind stress curl, so variables can be separated by defining

$$\psi_j = S_j(x,t)\cos(ly),$$
(46)

where $l = \pi/L$. Then (42) leads to

$$(S_{1xx} - l^2 S_1)_t + \beta S_{1x} + a_1^{-2}(S_2 - S_1)_t = -\frac{\tau_0 l}{\rho_0 H_1} + AD(S_1), \quad (47a)$$

$$(S_{2xx} - l^2 S_2)_t + \beta S_{2x} - a_2^{-2}(S_2 - S_1)_t = AD(S_2),$$
(47b)

where $D(S) = S_{xxxx} - 2l^2 S_{xx} + l^4 S$ is a dissipation term. Define the barotropic (depth-average) and baroclinic parts by

$$\bar{S} = \frac{(H_1 S_1 + H_2 S_2)}{H}, \quad \hat{S} = S_1 - S_2,$$
(48)

where $H = H_1 + H_2$. Then from (47)

$$(\bar{S}_{xx} - l^2 \bar{S})_t + \beta \bar{S}_x = -\frac{\tau_0 l}{\rho_0 H} + AD(\bar{S}), \qquad (49\text{a})$$

$$(\hat{S}_{xx} - l^2 \hat{S})_t + \beta \hat{S}_x - \frac{\hat{S}_t}{a^2} = -\frac{\tau_0 l}{\rho_0 H_1} + AD(\hat{S}), \qquad (49\text{b})$$

where a is the internal Rossby radius. Effectively, these are the rigid-lid barotropic and baroclinic Rossby wave equations, now with forcing and dissipation.

There is no flow across the eastern and western boundaries, so the zonal velocity component must be zero on each boundary in each layer, which is satisfied if the streamfunction is constant along each boundary: we choose $\psi_j = 0$ on the east and west boundaries.

With lateral dissipation, further conditions are required. The obvious condition is "no slip", for which $v_j = 0$ on the east and west boundaries, which requires

$$\psi_{jx} = 0 \quad \text{on } x = x_E, x_W. \qquad (50)$$

However, the quasigeostrophic equations do not properly represent small scale flow near a coastline, and in ocean modelling a common alternative is to use "free slip" conditions with $v_{jx} = 0$ on the east and west boundaries, which requires

$$\psi_{jxx} = 0 \quad \text{on } x = x_E, \ x_W. \qquad (51)$$

4.2. The steady state

First, consider the steady state that is obtained when the ocean has adjusted to reach equilibrium with the steady wind stress. From (47b), $S_2 = 0$ so after adjustment the lower layer is again at rest. From (47a),

$$S_{1x} = -C + \left(\frac{A}{\beta}\right) D(S_1), \qquad (52)$$

where for convenience we have defined a velocity scale $C = \tau_0 l / (\beta \rho_0 H_1)$.

For reasonable choices of parameters the length scale defined by $(A/\beta)^{1/3}$ is small compared to the scale L of the ocean basin, and dissipation effects are confined to thin layers near the lateral boundaries. Equation (52) can be re-written as

$$\epsilon^3 L^3 (S_{1xxxx} - 2l^2 S_{1xx} + l^4 S_1) - S_{1x} = C, \qquad (53)$$

where $\epsilon = (A/\beta)^{1/3}/L \ll 1$. This is a classic boundary layer problem that can be solved using matched asymptotic expansions. Away from the thin boundary layers, the "outer" equation is

$$S_{1x} = -C. \qquad (54)$$

Note that S_{1x} is the meridional flow v at $y = 0$, so in layer 1 in this example there is a southward geostrophic flow indirectly driven by the wind. (This geostrophic current is in balance with a zonal pressure gradient that is established during spin-up — see below.) The general solution of (54) is

$$S_1 = \alpha - Cx, \qquad (55)$$

where α is some constant to be determined by matching to the "inner" near-boundary behaviour.

4.2.1. *The western boundary layer*

With hindsight, for the "inner" expansion define the rescaled zonal coordinate X by $x = x_W + \epsilon X$, so (53) becomes

$$L^4 (S_{1xxxx} - 2l^2 \epsilon^2 S_{1xx} + l^4 \epsilon^4 S_1) - LS_{1x} = \epsilon CL. \qquad (56)$$

To leading order

$$L^4 S_{1xxxx} - LS_{1x} = 0. \qquad (57)$$

The general solution that does not grow exponentially eastward, and thus can be matched to the "outer" solution, is

$$S_1 = \gamma_0 + e^{-\frac{X}{2L}} \left[\gamma_1 \cos \left(\frac{\sqrt{3}X}{2L} \right) + \gamma_2 \sin \left(\frac{\sqrt{3}X}{2L} \right) \right]. \qquad (58)$$

A simple match of inner and outer solutions is sufficient: for large X (58) must match (55) as $x \to x_W$, so $\gamma_0 = \alpha - Cx_W$.

No slip conditions at the western boundary require $S_1 = S_{1x} = 0$ at $X = 0$, and thus

$$S_1 = \gamma_0 - \gamma_0 e^{-\frac{X}{2L}} \left[\cos \left(\frac{\sqrt{3}X}{2L} \right) + \frac{1}{\sqrt{3}} \sin \left(\frac{\sqrt{3}X}{2L} \right) \right]. \qquad (59)$$

If instead free slip conditions apply at the western boundary, then $S_1 = S_{1xx} = 0$ at $X = 0$, and

$$S_1 = \gamma_0 - \gamma_0 e^{-\frac{X}{2L}} \left[\cos\left(\frac{\sqrt{3}X}{2L}\right) - \frac{1}{\sqrt{3}} \sin\left(\frac{\sqrt{3}X}{2L}\right) \right]. \qquad (60)$$

4.2.2. The eastern boundary layer

For the "inner" expansion near the eastern boundary define instead the rescaled zonal coordinate X by $x = x_E - \epsilon X$. This time to leading order

$$L^4 S_{1xxxx} + L S_{1x} = 0 \qquad (61)$$

and to satisfy the boundary conditions at $X = 0$ and have no exponential growth toward the outer solution requires $S_1 = 0$.

Then a simple match to the outer solution (55) as $x \to x_E$ requires $\alpha = C X_E$, and hence $\gamma_0 = C(x_E - x_W) = CL$. (Note: thus to leading order there is no eastern boundary layer. The adjustments to satisfy the no slip boundary conditions are small higher order terms. With free slip conditions, the "outer" solution already satisfies the eastern boundary conditions and no boundary layer is required.)

4.2.3. The composite solution

A leading order additive composite asymptotic solution, valid in both the outer and boundary regions, can be found using the formula "inner + outer − overlap". The western "overlap" term is simply γ_0. For free slip conditions, the composite is

$$S_1 = C(x_E - x) - C(x_E - x_W)e^{-\frac{X}{2L}}$$
$$\times \left[\cos\left(\frac{\sqrt{3}X}{2L}\right) - \frac{1}{\sqrt{3}} \sin\left(\frac{\sqrt{3}X}{2L}\right) \right] \qquad (62)$$

with $X = (x - x_W)/\epsilon$, and there is an analogous formula for no slip conditions. Apart from details very near the western boundary, the free slip and no slip solutions are quite similar. In a narrow western boundary layer with width scale $(A/\beta)^{1/3}$ there is a relatively fast northward current with speed scale $1/\epsilon$ times that of the slow southward flow in the rest of the ocean.

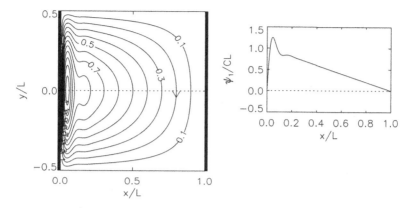

Fig. 3. Composite free-slip solution for the steady state upper layer streamfunction $\psi_1 = S_1(x)\cos(\pi y/L)$ (scaled by CL) in the ocean forced by wind stress. Streamlines (left panel) and $S_1(x)$ (right panel).

For a dissipation value of $A = 10^4 \text{m}^2\text{s}^{-1}$, together with the typical scales given previously, $(A/\beta)^{1/3}$ is about 80 km. For an ocean of width 50° of longitude at 40°N, the parameter ϵ is 0.0195. Using these as typical scales, Fig. 3 shows the composite free slip $S_1(x)$ on the right, and contours of the corresponding streamfunction $\psi_1(x, y)$ on the left. The wind stress drives a large-scale anticyclonic gyre with streamlines bunched up against the western boundary, where energy input by the wind is dissipated in a narrow region with strong gradients. This conceptual model helps us understand why the Gulf Stream occurs on the west side of the Atlantic Ocean.

From (25), the surface displacement is $\eta_1 = (f_0/g)S_1\cos(ly)$, while the interface displacement is $\eta_2 = -(g/g')\eta_1$. The surface displacement provides the horizontal pressure gradient in the upper ocean, and the interface displacement compensates such that there is zero gradient in layer 2.

Note: in the real world, the western boundary current separates from the coast at some point due to a combination of geography, nonlinearity, wind stress distribution, etc.: moreover, instabilities occur and a true steady state is never achieved.

4.3. *Spin-up of the stratified ocean*

It is instructive to see how the above steady state is reached starting from rest (a process known as "spin-up"), and how stratification influences this process (cf. Ref. 5). Suppose

$$S_1 = S_2 = \hat{S} = \bar{S} = 0 \text{ at time } t = 0. \tag{63}$$

4.3.1. *Initial adjustment*

Away from the eastern and western boundaries, the ocean initially evolves with no boundary influence. With wind stress curl independent of x, the x derivative terms are zero in the governing equations (47) and (49). Neglecting small dissipation terms,

$$\bar{S} = \frac{\tau_0 l}{\rho_0 H l^2} t, \quad \hat{S} = \frac{\tau_0 l a^2}{\rho_0 H_1 (1 + a^2 l^2)} t \ll \bar{S}. \tag{64}$$

The effect of the meridionally-varying wind on the thin surface Ekman layer is to create a meridional gradient in sea level that increases with time, and the resulting pressure gradient throughout the depth of the ocean leads to a corresponding zonal geostrophic flow within the ocean.

The presence of eastern and western boundaries has the effect of generating Rossby waves that propagate zonally, and the above initial adjustment is brought to an end principally by the arrival of these waves, with information travelling at the group velocity speed. The fastest Rossby waves are the long ($k \to 0$) westward-propagating waves. The fastest eastward-propagating waves are much slower, so most of the ocean is first influenced by the Rossby waves arriving from the eastern boundary. In the two-layer system, there are both barotropic and baroclinic modes to consider.

4.3.2. *Barotropic adjustment*

Consider some location $x = x_0$ between x_W and x_E, outside the western boundary region. With the rigid lid approximation, the fastest barotropic waves have group (and phase) velocity $-\beta/l^2$, and the time of arrival at x_0 from x_E is $t_0 = (x_E - x_0)l^2/\beta$. After this time barotropic adjustment ceases at x_0, and

$$\bar{S}(x_0) = C(x_E - x_0)\frac{H_1}{H} \text{ at time } t > t_0. \tag{65}$$

This is consistent with the barotropic part of the steady state solution described above. Near the western boundary, both short Rossby waves and dissipation play a role in developing the western boundary layer.

For typical scales, this barotropic adjustment takes just a few days. However, both the upper and lower layer flows are far from their eventual steady states at this stage, as the baroclinic mode takes much longer to "spin up".

4.3.3. *Baroclinic adjustment*

The baroclinic adjustment process is similar to barotropic adjustment, i.e., most of the ocean is adjusted by long baroclinic Rossby waves from the eastern boundary. However, the spin up time is much longer: the fastest baroclinic Rossby waves have zonal speed $-\beta a^2/(1 + a^2 l^2)$, and the spin-up time is several years for, say, the North Atlantic Ocean at mid-latitudes.

As the long baroclinic Rossby wave advances from the eastern boundary and reaches location x_0 at time $t_C = (x_E - x_0)(1 + a^2 l^2)/\beta a^2$, the baroclinic mode at that location equilibrates at

$$\hat{S}(x_0) = C(x_E - x_0) \quad \text{at time } t > t_C. \tag{66}$$

Again, this is consistent with the steady state solution above. After time t_C the lower layer is at rest between x_0 and x_E, and the upper layer has reached its steady state. As with the barotropic mode, the baroclinic western boundary layer also develops during the spin-up phase.

The whole process is best explained with a diagram, using typical scales as for the steady state. The panels in Fig. 4 show the streamfunctions ψ_1, ψ_2, $\bar{\psi}$ and $\hat{\psi}$ at various stages of the spin-up. At day 100 (dashed lines) the barotropic mode has reached its steady state over most of the ocean. The baroclinic mode is relatively small and zonally uniform except near the boundaries. In the individual layers, ψ_1 and ψ_2 are close to $\bar{\psi}$. The baroclinic mode then continues to increase

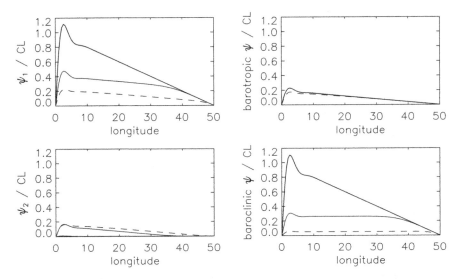

Fig. 4. Evolution of streamfunctions at $y = 0$, in the two-layer ocean forced by wind stress starting at day 0: day 100 (dashed), day 500 (solid) and day 2000 (thick lines). Left panels show ψ_1 (top) and ψ_2 (bottom); right panels show barotropic $\bar{\psi}$ (top) and baroclinic $\hat{\psi}$ (bottom), each scaled by CL.

linearly with time in the interior: by day 500 (thin solid line) the baroclinic Rossby wave from the western boundary (longitude 50°) has reached about longitude 40°. At this time $\bar{\psi}$ has changed little, ψ_2 has decreased and in particular is zero east of about longitude 40°, while ψ_1 has about doubled in size with the western boundary current increasing correspondingly. By day 2000 (thick line), the baroclinic long Rossby wave has crossed the ocean from the east, and the system has almost reached the equilibrium state with ψ_2 near zero and ψ_1 like the steady state in Fig. 3.

This example also serves to illustrate that baroclinic Rossby waves provide the ocean with a long adjustment time. In the real ocean, the density structure supports many baroclinic modes that are even slower than the one described above, so that time can be several decades. The baroclinic Rossby waves also provide the ocean with a long memory: the effects of wind stresses can have consequences at times and distances quite distant from the original events!

5. Topographic Effects and Multiple Equilibria

The "highs" and "lows" of mid-latitude weather systems fluctuate from day to day, with the systems generally moving eastward, in part carried along by the prevailing westerly mean zonal flow. In some circumstances, the mean flow weakens and a "block" develops that slows down and diverts weather systems: it seems that the system may have different "regimes". One of several possible mechanisms for "blocking" involves interaction between the mean flow, Rossby waves and topography. Conceptual models were put forward by Charney & DeVore[6] and Hart[7] using a single-layer beta-plane with a zonally periodic channel notionally representing a mid-latitude strip around the Earth. In this section this mechanism is described using a similar beta-plane model to that in Davey.[8]

5.1. *Formulation of a periodic channel model*

Consider a system with a single layer of mean depth H. The configuration is a zonally periodic channel with length $2\pi L$ and northern and southern boundaries ("sidewalls") located at $y=\pm\pi L/2$ for simplicity. There is topography $h(x, y)$ at the bottom, and a flat upper surface, and horizontal diffusivity is neglected.

The quasigeostrophic potential vorticity equation that governs flow outside of thin boundary layers is, from (23) and (26c),

$$\frac{D_g}{Dt}\left(\nabla^2\psi + \beta y + \left(\frac{f_0}{H}\right)h\right) = \left(\frac{f_0}{H}\right)(w_{ET} - w_{EB}). \qquad (67)$$

Suppose the upper surface moves at a prescribed velocity $\underline{u}_T$: it is this movement that drives flow in the system. Through viscous effects in the thin Ekman layers at the top and bottom boundaries it is found that

$$w_{ET} - w_{EB} = E^{\frac{1}{2}}\frac{H(\zeta_T - 2\zeta)}{2}, \qquad (68)$$

where ζ_T is the upper surface vorticity, and $E = 2\nu/f_0H^2 \ll 1$ is the non-dimensional Ekman number. (Derivations of this expression may be found in textbooks. Note that $E^{1/2}H$ is a scale for the vertical extent of the Ekman layer.) This setup is similar to that used in laboratory experiments with rotating tanks which have differentially rotating lids.

The following general circulation condition applies, stated here without proof:

$$\oint \underline{u}_t + f_0 E^{\frac{1}{2}} \left(\underline{u} - \frac{\underline{u}_T}{2} \right) \cdot \underline{dl} = 0, \tag{69}$$

where the path integral is taken around a closed streamline of the geostrophic flow. In the periodic channel, this condition includes zonally periodic streamlines.

In particular, suppose $\underline{u}_T = (2U_0, 0)$, with constant U_0. Then (67) is

$$\frac{D_g}{Dt} \left(\nabla^2 \psi + \beta y + \left(\frac{f_0}{H} \right) h \right) = -f_0 E^{1/2} \zeta. \tag{70}$$

With flat topography $h = 0$ there is a uniform geostrophic zonal flow $\underline{u} = (U_0, 0)$: notionally this corresponds to an atmospheric zonal flow associated with a meridional pressure gradient and a meridional temperature gradient.

Define zonal and meridional averages by

$$\langle \cdots \rangle = \frac{1}{2\pi L} \int\limits_{-\pi L}^{\pi L} \ldots dx \quad \text{zonal average}, \tag{71a}$$

$$[\ldots] = \frac{1}{\pi L} \int\limits_{0}^{\pi L} \ldots dy \quad \text{meridional average}. \tag{71b}$$

Then the geostrophic streamfunction can be written as

$$\psi = -U(t)y + \Phi(y, t) + \phi(x, y, t), \tag{72}$$

where $U = [\langle u \rangle]$ is the mean zonal geostrophic flow, Φ is the streamfunction for the zonal mean shear $\langle u \rangle - U$, and ϕ is the streamfunction for wave-like flow. By construction, $\langle \phi \rangle = 0$ and $[\Phi_y] = 0$.

The streamfunction ψ is constant along the sidewalls: choose

$$\psi = -\frac{U\pi L}{2} \text{ on } y = \frac{\pi L}{2}, \quad \psi = \frac{U\pi L}{2} \text{ on } y = -\frac{\pi L}{2} \tag{73}$$

and $\Phi = \phi = 0$ on the sidewalls.

Equation (70) can be re-written as

$$\Phi_{yyt} + \nabla^2\phi_t + \langle u \rangle \left(\nabla^2\phi + \left(\frac{f_0}{H} \right) h \right)_x + \phi_x(\beta + \Phi_{yyy})$$

$$+ J\left(\phi, \nabla^2\phi + \left(\frac{f_0}{H} \right) h \right) = -f_0 E^{\frac{1}{2}}(\Phi_{yy} + \nabla^2\phi), \qquad (74)$$

where J is the Jacobian operator. Taking the zonal average of (74) leads to

$$\Phi_{yyt} + \left\langle J\left(\phi, \nabla^2\phi + \left(\frac{f_0}{H} \right) h \right) \right\rangle + f_0 E^{\frac{1}{2}}\Phi_{yy} = 0. \qquad (75)$$

Using $\langle J(\phi, h) \rangle = -\langle \phi h_x \rangle_y$, and a similar result for $\langle J(\phi, \nabla^2\phi) \rangle$, (75) can be expressed as

$$\left(\Phi_{yt} - \left\langle \phi \left(\nabla^2\phi + \left(\frac{f_0}{H} \right) h \right)_x \right\rangle + f_0 E^{\frac{1}{2}}\Phi_y \right)_y = 0. \qquad (76)$$

Integrating (76) with respect to y from the sidewall at $y = -\pi L/2$ leads to

$$-\Phi_{yt} + \left\langle \phi \left(\nabla^2\phi + \left(\frac{f_0}{H} \right) h \right)_x \right\rangle - f_0 E^{\frac{1}{2}}\Phi_y$$

$$= (-\Phi_{yt} - f_0 E^{\frac{1}{2}}\Phi_y)|_{y=-\frac{\pi L}{2}}. \qquad (77)$$

From (69),

$$U_t - \Phi_{yt} + f_0 E^{\frac{1}{2}}(U - \Phi_y - U_0) = 0 \quad \text{at } y = -\frac{\pi L}{2}. \qquad (78)$$

Thus,

$$U_t - \Phi_{yt} + \left\langle \phi \left(\nabla^2\phi + \left(\frac{f_0}{H} \right) h \right)_x \right\rangle - f_0 E^{\frac{1}{2}}\Phi_y + f_0 E^{\frac{1}{2}}(U - U_0) = 0. \qquad (79)$$

Taking the meridional average of (79), using $\Phi = 0$ on the sidewalls and $[\langle \phi \nabla^2 \phi_x \rangle] = 0$ (from periodicity and $\phi = 0$ on sidewalls), leads to

$$U_t + \left(\frac{f_0}{H} \right) [\langle \phi h_x \rangle] + f_0 E^{\frac{1}{2}}(U - U_0) = 0. \qquad (80)$$

This is the key equation, and it has a simple interpretation. Recalling that the streamfunction is proportional to pressure perturbations, the second term is proportional to the drag on the flow exerted by the pressure field associated with the wave-like flow component. Because ϕ depends on U, the drag also depends on U.

At this stage, the model is nonlinear and must generally be solved numerically to determine U, Φ and ϕ.

5.2.   Steady quasilinear flow

Analytic progress can be made by neglecting wave-wave interactions in the quasigeostrophic potential vorticity equation, and by considering steady flow. For algebraic simplicity, the mean shear component Φ is also neglected here. Effectively, we set $\langle u \rangle = U$ and $J(\Phi + \phi, \Phi_{yy} + \nabla^2\phi + (f_0/H)h) = 0$ in (74), to obtain

$$U \left(\nabla^2\phi + \left(\frac{f_0}{H}\right)h \right)_x + \beta\phi_x + f_0 E^{\frac{1}{2}}\nabla^2\phi = 0. \tag{81}$$

Consider topography $h(x, y)$ of the form

$$h = h_0 \cos(ly) \sum_{m=1}^{M} F_m \cos(kx), \tag{82}$$

where $l = 1/L$, $k = m/L$ and h_0 is a topography height scale. Given this form for h, we can similarly express ϕ in the form

$$\phi = \cos(ly) \sum_{m=1}^{\infty} A_m \cos(kx) + B_m \sin(kx) \tag{83}$$

and obtain from (81) expressions for the Fourier coefficients A_m and B_m:

$$A_m = \frac{U f_0 k^2 (U K^2 - \beta)\left(\frac{h_0}{H}\right) F_m}{k^2(U K^2 - \beta)^2 + f_0^2 E K^4}, \tag{84a}$$

$$B_m = \frac{-U f_0 E^{\frac{1}{2}} K^2 \left(\frac{h_0}{H}\right) k F_m}{k^2(U K^2 - \beta)^2 + f_0^2 E K^4}, \tag{84b}$$

where $K^2 = k^2 + l^2$. These expressions can be used in (80) to obtain an equation for U. Noting that $[\langle \phi h_x \rangle] = -(h_0/4)\sum_m kB_m F_m$, this leads to

$$U = U_0 - \left(\frac{U}{4}\right) f_0^2 \left(\frac{h_0}{H}\right)^2 \sum_m \frac{k^2 K^2 F_m^2}{k^2(UK^2 - \beta)^2 + f_0^2 E K^4} \qquad (85)$$

which is an odd-order polynomial equation for U. Using $k = m/L$ and $l = 1/L$ and some re-arrangement, this is equivalent to

$$\hat{U} = 1 - \left(\frac{\hat{U}}{4}\right)\left(\frac{h_0}{H}\right)^2 \sum_m \frac{m^2(1 + m^2) F_m^2}{m^2\left(\hat{U} R_0(1 + m^2) - \hat{\beta}\right)^2 + E(1 + m^2)^2}. \qquad (86)$$

Thus, $\hat{U} = U/U_0$ can be determined given the non-dimensional parameters $R_0 = U_0/f_0 L$ (Rossby number), $\hat{\beta} = \beta L/f_0$, E, h_0/H and the topography shape coefficients F_m. From (86) it can be shown that $0 < \hat{U} \le 1$, i.e., the topographic drag always slows down the mean zonal flow.

Note that (86) is a nonlinear equation for $\hat{U}$, but it can be regarded as a linear equation for R_0 given $\hat{U}$ (i.e., for U_0 given U), which is convenient for calculating solutions.

5.3.   *Blocked and unblocked flows*

The combination $UK^2 - \beta$ has an important role. The physical interpretation is that $U - \beta/K^2$ is the Doppler-shifted phase speed of free barotropic rigid-lid Rossby waves with wavenumber K in a mean zonal flow of speed U. When U is positive, this speed may be close to zero for particular K.

To describe this simply, consider topography with just one zonal wavenumber, so the wave-like flow has the same wavenumber and $\nabla^2 \phi = -K^2 \phi$. The quasilinear quasigeostrophic vorticity equation (81) becomes

$$(\beta - UK^2)\phi_x - f_0 E^{\frac{1}{2}} K^2 \phi + U\left(\frac{f_0}{H}\right) h_x = 0. \qquad (87)$$

For small $E^{1/2}$ and $UK^2 - \beta$ not close to zero, the dominant balance is between vorticity generated by the flow over the topography and advection of vorticity by the mean flow (modified by the Rossby wave), giving

$$\phi \approx \frac{Uf_0}{H(UK^2 - \beta)} h. \tag{88}$$

In this case, ϕ is proportional to h, the topographic drag is low, and $U/U_0 \approx 1$: the mean flow is "unblocked". (Note that ϕ may have the same or opposite sign to h, depending on the sign of $UK^2 - \beta$.)

If instead $UK^2 - \beta$ is close to zero, the dominant balance is between vorticity generated by the flow over the topography and dissipation, giving

$$\phi \approx \frac{U}{E^{\frac{1}{2}}K^2 H} h_x. \tag{89}$$

Now the wave-like flow is out of phase with h, resulting in larger drag and a reduction in U. For weak dissipation, this balance requires a large amplitude wave and substantial reduction in U, and the flow is "blocked".

5.4. *Multiple equilibria and regimes*

Another feature of this simple model is that (85) may have more than one solution for the same parameter choices. This is best illustrated by an example, using topography constructed with ten Fourier modes with shape shown in Fig. 5, and parameters $E = 10^{-4}$, $h/h_0 = 0.2$ and $\beta L/f_0 = 0.2$.

Figure 6 shows U/U_0 as a function of $R_0 = U_0/f_0 L$, which may be regarded as showing how U/U_0 depends on the driving term U_0. For several ranges of values of U_0, there are three values of U/U_0. The middle value is unstable, and in a time-dependent model the solution would be attracted toward one of the other two values. The lower value corresponds to a blocked flow, while the other is unblocked. The various ranges of U_0 with multiple equilibria correspond to various zonal wavenumbers m for which the Rossby wave with that wavenumber may resonate and generate large topographic drag, with the range furthest to the right in Fig. 6 corresponding to $m = 1$.

The two stable equilibria at $R_0 = 0.2$ are illustrated in Fig. 7. The upper panel shows streamlines (contours of the streamfunction)

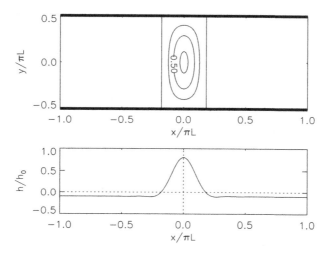

Fig. 5. The topography shape $h(x, y)/h_0$ for the multiple equilibria example: plan view (top) and profile along $y = 0$ (bottom).

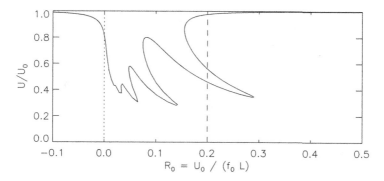

Fig. 6. U/U_0 as a function of $U_0/f_0 L$ for the multiple equilibria example.

for the unblocked option: the flow is only slightly disturbed by the topography, and $U/U_0 = 0.97$. By contrast the lower panel shows the blocked alternative: the zonal mean is reduced to $U/U_0 = 0.46$, the wave amplitude is much larger, and flow is sufficiently distorted to produce closed streamlines within the channel. Similarly, there are two stable equilibria at $R_0 = 0.1$ with flow patterns (not shown) dominated by zonal wavenumber 2.

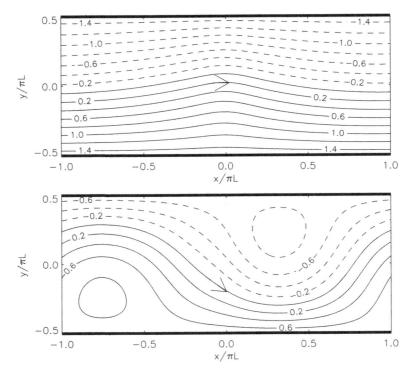

Fig. 7.   Contours of the streamfunctions for the stable equilibria at $R_0 = 0.2$ in Fig. 6. Top: "unblocked" flow with $U/U_0 = 0.97$. Bottom: "blocked" flow with $U/U_0 = 0.46$.

Figure 6 also illustrates how the flow may be locked into a blocked or unblocked regime. Consider starting at the blocked state with $R_0 = 0.2$. As the driving value U_0 is increased the flow remains blocked until $R_0 \approx 0.3$: beyond this point the system suddenly adjusts to an unblocked state as the only possibility, and remains unblocked if U_0 is increased further. But if U_0 is then decreased, the system remains in the unblocked regime until $R_0 \approx 0.16$, beyond which point the system switches to a more blocked state dominated by $m = 2$.

While a zonal channel may seem far removed from the spherical Earth, atmospheric flows may form waveguides for Rossby waves and studies such as that of Ref. 9 indicate that the mechanism of wave-topography-mean-flow interaction inducing multiple regimes is relevant to the Earth's atmosphere.

6. Summary

The focus in the preceding sections has been on quasigeostrophic dynamics, applicable outside tropical regions. The examples demonstrate how idealised scenarios with simplified sets of equations can be formulated, and solutions can be found that help us understand the processes underlying various oceanic and atmospheric phenomena.

Note that uses of geophysical fluid dynamics are not limited to areas where geostrophic balances prevail. Rotational effects remain important in the tropics, and for example the "equatorial beta plane" approximation provides insight into the important effects of equatorially-trapped waves. Nor are they limited to the Earth: there are abundant examples of applications to fluid flows on other planets.

Exercises

Rossby waves and ocean spin-up

Consider the unforced zonal wave equation for baroclinic Rossby waves with meridional wavenumber l:

$$\hat{S}_t - \beta a^2 \hat{S}_x - a^2 (\hat{S}_{xx} - l^2 \hat{S})_t = 0.$$

Derive expressions for the fastest eastward and westward group velocities. In an ocean of width L, deduce that the Rossby waves emanating from the east and west boundaries first meet a distance $L/9$ from the western side.

    Answer: fastest eastward $\partial\omega/\partial k = -\beta a^2/(1 + a^2 l^2)$; fastest westward $\partial\omega/\partial k = -\beta a^2/8(1 + a^2 l^2)$.

Rossby waves and topography

Consider steady zonally periodic flow in a beta-plane channel with length $2\pi L$, width πL, mean depth H and topography $h =$

$h_0 \cos(kx)\cos(ly)$ with $k = l = 1/L$. The geostrophic streamfunction is $\psi = -Uy + \phi(x,y)$, with governing equation

$$U\left(\nabla^2\phi + \left(\frac{f_0}{H}\right)h\right)_x + \beta\phi_x + f_0 E^{\frac{1}{2}}\nabla^2\phi = 0.$$

Quasilinear theory requires

$$\left(\frac{f_0}{H}\right)[<\phi h_x>] + f_0 E^{\frac{1}{2}}(U - U_0) = 0,$$

where [] and <> denote averages across and along the channel respectively, and U_0 is the zonal flow when $h_0 = 0$. Derive the equation

$$2R_0 = 2R + \frac{R(h_0/H)^2}{((2R - \hat\beta)^2 + 4E)},$$

where $R = U/f_0 L$, $R_0 = U_0/f_0 L$, $\hat\beta = \beta L/f_0$.

Consider $E = 0.001$, $\hat\beta = 0.2$. By plotting R_0 as a function of R for increasing values of h_0/H, or otherwise, find the value of h_0/H for which multiple equilibria first occur.

Answer: 0.0673

Energy conservation

For a two-layer β-plane rigid lid system, the linearised free quasi-geostrophic equations are

$$\nabla^2\psi_{jt} + (3 - 2j)\left(\frac{f_0}{H_j}\right)\eta_t + \beta\psi_{jx} = 0$$

for $j = 1, 2$, where $\eta = \left(\frac{f_0}{g'}\right)(\psi_2 - \psi_1)$. Show that

$$\left(\sum_1^2 H_j(u_j^2 + v_j^2) + g'\eta^2\right)_t$$

$$= 2\sum_1^2 H_j((\psi_j\psi_{jxt})_x + (\psi_j\psi_{jyt})_y + \left(\frac{\beta}{2}\right)(\psi_j^2)_x),$$

where $u_j = -\psi_{jy}$ and $v_j = \psi_{jx}$.

Consider flow in a zonally periodic channel with free slip northern and southern sides. Deduce that $[< \sum_1^2 H_j(u_j^2 + v_j^2) + g'\eta^2 >]$ is constant, where $[\,]$ and $<>$ are averages as defined in the previous exercise. (This is proportional to the sum of kinetic and potential energy in the channel.) Note: you may assume that $<\psi_{jyt}> = 0$ on the sidewalls.

The result is also true for the nonlinear version of the quasi-geostrophic equations: prove this as an extension to this exercise.

References

1. A. E. Gill, *Atmosphere-Ocean Dynamics*, Academic Press, New York, 1982.
2. J. Pedlosky, *Geophysical Fluid Dynamics*, 2nd edn., Springer, New York, 1992.
3. G. K. Vallis, *Atmospheric and Oceanic Fluid Dynamics*, Cambridge University Press, Cambridge, 2006.
4. W. H. Munk, On the wind-driven ocean circulation, *J. Meteorol.* **7** (1950), 79–93.
5. D. L. T. Anderson and A. E. Gill, Spin-up of a stratified ocean, with applications to upwelling, *Deep-Sea Research* **22**, (1973), 583–596.
6. J. G. Charney and J. G. DeVore, Multiple flow equilibria in the atmosphere and blocking, *J. Atmos. Sci.* **36**, (1979), 1205–1216.
7. J. E. Hart, Barotropic quasi-geostrophic flow over anisotropic mountains, *J. Atmos. Sci.* **36**, (1979), 1736–1746.
8. M. K. Davey, A quasi-linear theory for rotating flow over topography. Part 1. Steady β-plane channel, *J. Fluid Mech.* **99** (1980), 267–292.
9. S. Yang, B. Reinhold and E. Källén, Multiple weather regimes and baroclinically forced spherical resonance, *J. Atmos. Sci.* **54**, (1997), 1397–1409.

Chapter 2

Multiple Deck Theory

S. N. Timoshin

Department of Mathematics, UCL, Gower Street,
London WC1E 6BT, UK,
s.timoshin@ucl.ac.uk

This chapter is an introduction to the asymptotic theory of laminar viscous flows at large Reynolds numbers. Starting with a derivation of the governing Navier–Stokes equations, the theory is presented as a sequence of worked examples of increasing complexity including the classical boundary layer, the triple-deck flow with viscous-inviscid interaction and the double-deck interactive flow in a wall jet, briefly touching upon supersonic and condensed flow regimes. The theory is used to study the phenomena of upstream influence, flow induced by a local roughness, short-scale boundary layer separation near an irregularity in the wall geometry, self-induced separation as part of a global picture, and instability in the flow.

1. Introduction

Fluid flows with large Reynolds numbers are encountered in almost every application, from meteorology to aeronautics and human physiology. Theoretical developments in these areas rely on a combination of computation and asymptotic analysis, the latter being the subject of this chapter. In asymptotic calculations, the flow structure can be visualised as consisting of several layers or decks, hence the multiple-deck theory. The reader is expected to be familiar with the method of matched asymptotic expansions (as described elsewhere in this volume[1]) but, since the governing Navier–Stokes equations are derived in Sec. 2 below, only a passing familiarity with fluid dynamics should be sufficient to understand the mathematical aspects of the theory.

The asymptotic analysis of the flow past a simple geometry begins in Sec. 3 with two complementary limit solutions — the inviscid Euler flow and the Prandtl boundary layer. In Sec. 4, the triple-deck flow structure is introduced, bringing in the notion of viscous-inviscid interaction formalised in a very non-trivial boundary-value problem. Studying the properties of the triple-deck flow in Sec. 5, we encounter flow separation (both localised and large-scale), and growing waves which are identified as the lower-branch Tollmien-Schlichting instability. In Sec. 6, further examples of multi-deck flows are mentioned to introduce the phenomenon of upstream influence and to give further illustration to self-induced processes in boundary layer flows.

The entire area has been reviewed extensively[2–5] and a curious reader will undoubtedly find the field of asymptotic fluid dynamics fascinating and rewarding, especially in the topics which are beyond the scope of this chapter (such as existence/uniqueness of solutions, absolute/convective instability, nonlinear waves, bifurcations and chaos, and numerous new applications). The introductory level of the following discourse should be sufficient to equip the reader with a new powerful tool of applied analysis, not limited to the area of fluid dynamics.

2.   Incompressible Viscous Fluid Flow: The Navier–Stokes Equations

This section gives a brief derivation of the Navier–Stokes equations for the flow of an incompressible viscous fluid, see e.g., Refs. 6, 7 for more detail.

The fluid is treated as a continuum characterised by such quantities as density, velocity, pressure and so on. In the so-called Lagrangian specification, the flow parameters are studied following the motion of fluid particles, i.e., infinitesimal volumes of fluid. Alternatively, in the Eulerian flow specification, all flow quantities are regarded as functions of space and time without an explicit reference to flow particles. For instance, in the Cartesian coordinates with the unit vectors $\mathbf{e}_i, i = 1, 2, 3$, we then have the Eulerian velocity field in the flow, $\mathbf{u} = (u_1, u_2, u_3) = u_i \mathbf{e}_i$, as a function of time t and the position vector, $\mathbf{r} = (x_1, x_2, x_3) = x_i \mathbf{e}_i$, so that $\mathbf{u} = \mathbf{u}(\mathbf{r}, t)$. Note the summation convention over repeated indices in the above expressions. The conservation laws in fluid dynamics are more transparent in the Lagrangian specification (i.e., following fluid particles), however, the resulting equations of motion are easier to handle in the Eulerian specification.

The material derivative. Consider any scalar quantity associated with the flow, for example the density $\rho = \rho(\mathbf{r}, t)$, written in the Eulerian specification. Let $\mathbf{r} = \mathbf{r}(\mathbf{r}_0, t)$ be the trajectory of a particle whose initial position vector is $\mathbf{r}_0$. Since the velocity of the particle is $\mathbf{u} = \partial \mathbf{r}(\mathbf{r}_0, t)/\partial t$, the rate of change of density in the chosen particle is given by

$$\frac{\partial \rho(\mathbf{r}(\mathbf{r}_0, t), t)}{\partial t} = \frac{\partial \rho(\mathbf{r}, t)}{\partial t} + (\mathbf{u} \cdot \nabla) \rho(\mathbf{r}, t) = \frac{D\rho(\mathbf{r}, t)}{Dt}. \qquad (2.1)$$

Here, $\nabla = \mathbf{e}_i \partial/\partial x_i$ and the notation $D/Dt = \partial/\partial t + (\mathbf{u} \cdot \nabla)$ is used for the derivative following a fluid particle or the *material* derivative. In this context, the partial time derivative in (2.1), $\partial/\partial t$, is called the *local* time derivative and the operator $(\mathbf{u} \cdot \nabla)$ is the *convective* derivative. The formula applies to vector fields as well giving, for instance, the acceleration of a fluid particle as $\mathbf{a} = D\mathbf{u}/Dt = \partial\mathbf{u}/\partial t + (\mathbf{u} \cdot \nabla) \mathbf{u}$.

Conservation of mass. A material volume is a part of the flow containing the same fluid particles at all times. The statement of mass conservation in a small material volume, δV, may be written as

$$\frac{D(\rho \delta V)}{Dt} = \delta V \left[\frac{D\rho}{Dt} + \rho \frac{1}{\delta V} \frac{D(\delta V)}{Dt} \right] = 0. \qquad (2.2)$$

Note that

$$\lim_{\delta V \to 0} \frac{1}{\delta V} \frac{D(\delta V)}{Dt} = \frac{\partial u_i}{\partial x_i} = \operatorname{div} \mathbf{u}. \qquad (2.3)$$

The limit $\delta V \to 0$ applied to (2.2) yields

$$\frac{\partial \rho}{\partial t} + (\mathbf{u} \cdot \nabla)\,\rho + \rho\,\mathrm{div}\mathbf{u} = 0. \tag{2.4}$$

The fluid is incompressible if $D(\delta V)/Dt = 0$ leading to the so-called continuity condition,

$$\mathrm{div}\mathbf{u} = 0. \tag{2.5}$$

For a homogeneous ($\rho \equiv \mathrm{const}$) incompressible fluid mass conservation holds if the velocity field satisfies the continuity equation (2.5).

The momentum equation. The second Newton's law of mechanics relates the momentum changes in a body to the total force acting on that body. With application to the fluid contained in a material volume $V(t)$ with surface $A(t)$, we distinguish between long-range forces (using here gravity as an example) and short-range forces due to molecular interactions taking place at the body surface (pressure and friction). The momentum equation can then be written as

$$\int\limits_{V(t)} \rho \frac{D\mathbf{u}}{Dt} dV = \int\limits_{V(t)} \rho \mathbf{g} dV + \int\limits_{A(t)} \mathbf{\Sigma} dA, \tag{2.6}$$

where $\mathbf{g}$ is the gravitational acceleration. The surface integral is obtained by writing the local force, $d\mathbf{F}$, acting on an area element, dA, in terms of the stress vector, $\mathbf{\Sigma}$, defined by $d\mathbf{F} = \mathbf{\Sigma} dA$. Intuitively, we expect the stresses to be related to deformations of fluid elements — pressure forces tend to compress fluid particles, whereas friction arises when liquid layers slide relative to each other. Formally, both stresses and deformations in a moving medium are characterised by tensor quantities as follows.

At a fixed moment in time, consider two particles with the position vectors $\mathbf{r}$ and $\mathbf{r} + \delta\mathbf{r}$ separated by a small distance $\delta\mathbf{r} = \mathbf{e}_i \delta x_i$. Then, in the Taylor expansion, $\mathbf{u}(\mathbf{r} + \delta\mathbf{r}) = \mathbf{u}(\mathbf{r}) + \mathbf{e}_i \delta x_j \partial u_i(\mathbf{r})/\partial x_j + \cdots$, we define two second-order tensors,

$$e_{ij} = \frac{1}{2}\left(\frac{\partial u_i}{\partial x_j} + \frac{\partial u_j}{\partial x_i}\right), \ \xi_{ij} = \frac{1}{2}\left(\frac{\partial u_i}{\partial x_j} - \frac{\partial u_j}{\partial x_i}\right), \tag{2.7}$$

and hence write the local velocity field as

$$\mathbf{u}\left(\mathbf{r}+\delta\mathbf{r}\right) = \underbrace{\mathbf{u}\left(\mathbf{r}\right)}_{\text{translation}} + \underbrace{\xi_{ij}\delta x_j \mathbf{e}_i}_{\text{rotation}} + \underbrace{e_{ij}\delta x_j \mathbf{e}_i}_{\text{deformation}} + \text{higher order terms.}$$

(2.8)

The first term in (2.8) corresponds to a pure translation with constant speed. It is easy to verify that the second term, with the antisymmetric tensor ξ_{ij}, gives the velocity field identical to that in a rigid body rotation with the angular velocity $\mathrm{curl}(\mathbf{u})/2$. The third term represents deformation characterised by the symmetric tensor e_{ij} known as the rate-of-strain tensor. Translation and rotation can be maintained without extra forces; therefore short-range forces should be connected with the fluid deformation.

Returning now to the surface integral in (2.6), one can notice that the stress vector, $\boldsymbol{\Sigma}$, depends on the orientation of the surface element. However, this dependence is not arbitrary. Indeed, let us assume that the typical linear size of the fluid volume, δr say, is small. Then the volume integrals in (2.6) are of order $O((\delta r)^3)$ whereas the surface integral is $O((\delta r)^2)$. The balance of terms requires then

$$\int_{A(t)} \boldsymbol{\Sigma} dA = 0. \tag{2.9}$$

From the last relation, it can be shown that the stress vector is linearly related with the outward unit normal $\hat{\mathbf{n}}$ of the area dA. More precisely, if we write in components $\boldsymbol{\Sigma} = \Sigma_i \mathbf{e}_i$, $\hat{\mathbf{n}} = n_i \mathbf{e}_i$ then Σ_i can be written in terms of a second-order stress tensor, σ_{ij}, as $\Sigma_i = \sigma_{ji} n_j$ or $\Sigma_i = \sigma_{ij} n_j$. We have used here the fact (which is intuitively expected but also follows rigorously from the conservation of the angular momentum) that the stress tensor is symmetric, $\sigma_{ij} = \sigma_{ji}$. Substituting the stress vector $\boldsymbol{\Sigma}$ into (2.6) and applying the divergence theorem we have, for the i-component,

$$\rho\left[\frac{\partial u_i}{\partial t} + (\mathbf{u} \cdot \nabla) u_i\right] = \rho g_i + \frac{\partial \sigma_{ij}}{\partial x_j}. \tag{2.10}$$

This relation is known as the Cauchy equation, applicable to a general continuous medium.

The constitutive relation. For an incompressible Newtonian fluid, we assume a linear relation between the stress tensor and the rate-of-strain tensor,

$$\sigma_{ij} = -p\delta_{ij} + 2\mu e_{ij}, \tag{2.11}$$

where δ_{ij} is the Kronecker tensor, p is pressure and μ is the viscosity coefficient. In practice, the viscosity coefficient changes with temperature, however here we take μ to be constant. For an inviscid fluid we have the pressure as the only active stress, $\sigma_{ij} = -p\delta_{ij}$. The formula (2.11) can be derived as a general linear relation between second-order tensors σ_{ij} and e_{ij} with the additional requirement of isotropy, that is, independence of the flow properties from the orientation of the coordinate axes.

The Navier–Stokes equations. Substitution of (2.11) into (2.10) together with the expression for e_{ij} in (2.7) leads to the Navier–Stokes equations of motion for a homogeneous incompressible viscous fluid,

$$\rho \left[\frac{\partial \mathbf{u}}{\partial t} + (\mathbf{u} \cdot \nabla) \mathbf{u} \right] = \rho \mathbf{g} - \nabla p + \mu \nabla^2 \mathbf{u}, \tag{2.12}$$

$$\mathrm{div}\,\mathbf{u} = 0, \tag{2.13}$$

where the continuity condition is included as a consequence of mass conservation. In what follows, we omit the gravity term which can be incorporated into the modified pressure, $p^* = p + \rho f$, where f is the gravitational potential such that $\mathbf{g} = -\nabla f$.

For a viscous fluid, we adopt the no-slip condition on any solid boundary insisting on an exact match between the flow velocity and the velocity of the boundary. In particular, $\mathbf{u} = 0$ on the boundary of a body at rest. In an inviscid flow, only the normal component of the fluid velocity needs to match the velocity of the solid boundary, allowing slip along the walls.

The Reynolds number. Suppose a solid body is immersed in a uniform stream of incompressible viscous fluid. For definiteness, it may be a sphere of radius L, placed in a current with constant pressure p_∞ flowing along the x_1-axis with speed u_∞. The equations of motion are (2.12), (2.13) with the $\mathbf{g}$ term omitted. The boundary conditions in the far field are $\mathbf{u} \to u_\infty \mathbf{e}_1$, $p \to p_\infty$ as $|\mathbf{r}| \to \infty$. If Γ denotes the body surface then we also have the no-slip conditions,

$\mathbf{u} = 0$ if $\mathbf{r} \in \mathbf{\Gamma}$. There are five parameters in the problem formulation: the fluid viscosity μ and density ρ, the speed and pressure in the incoming flow, u_∞ and p_∞ and the length parameter of the body, L. After a change to non-dimensional variables (denoted with an overbar) according to $\mathbf{r} = L\bar{\mathbf{r}}$, $t = (L/u_\infty) \bar{t}$, $\mathbf{u} = u_\infty \bar{\mathbf{u}}$, $p = p_\infty + \rho u_\infty^2 \bar{p}$, the problem formulation becomes:

$$\frac{\partial \bar{\mathbf{u}}}{\partial \bar{t}} + \left(\bar{\mathbf{u}} \cdot \overline{\nabla}\right) \bar{\mathbf{u}} = -\overline{\nabla}\bar{p} + \frac{1}{\mathrm{Re}}\overline{\nabla}^2 \bar{\mathbf{u}}, \quad \overline{\nabla} \cdot \bar{\mathbf{u}} = 0, \tag{2.14}$$

$$\bar{\mathbf{u}} \to \mathbf{e}_1, \quad \bar{p} \to 0 \text{ as } \bar{\mathbf{r}} \to \infty; \quad \bar{\mathbf{u}} = 0 \text{ if } \bar{\mathbf{r}} \in \overline{\mathbf{\Gamma}}. \tag{2.15}$$

The non-dimensional flow field is fully determined by just one parameter, the Reynolds number,

$$\mathrm{Re} = \frac{\rho u_\infty L}{\mu} = \frac{u_\infty L}{\nu}. \tag{2.16}$$

The quantity $\nu = \mu/\rho$ appearing in the last expression is the kinematic viscosity of the fluid. In this chapter, we are mostly concerned with a two-dimensional (2D) flow in the (x, y)-plane with the velocity components (u, v) where, omitting the overbars, the non-dimensional Navier–Stokes equations (2.14) can be written as,

$$u_t + uu_x + vu_y = -p_x + \mathrm{Re}^{-1}(u_{xx} + u_{yy}), \tag{2.17}$$

$$v_t + uv_x + vv_y = -p_y + \mathrm{Re}^{-1}(v_{xx} + v_{yy}), \tag{2.18}$$

$$u_x + v_y = 0. \tag{2.19}$$

3. Flow Past a Flat Plate at a Large Reynolds Number

We now proceed to derive a few solutions of the Navier–Stokes equations (2.17)–(2.19) at large Reynolds numbers. Consider first the flow past a semi-infinite flat plate aligned with the x-axis, Fig. 1(a). This is a classical problem in the boundary layer theory.[8] In the far field upstream, the flow is assumed parallel to the plate with constant speed,

$$u \to 1, \ v \to 0, \ p \to 0 \text{ as } x^2 + y^2 \to \infty, \tag{3.1}$$

and the no-slip conditions on the solid wall become,[a]

$$u = v = 0 \text{ at } y = 0 \text{ for } x \geq 0. \tag{3.2}$$

---

[a]There seems to be an issue with the far-field conditions downstream, especially close to the wall, but it proves irrelevant in the solution obtained.

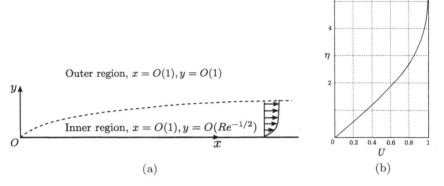

Fig. 1.   Flow past a flat plate (a) Outer and inner flow regions, (b) Blasius velocity profile.

3.1.   *Outer limit — inviscid flow*

Let $\mathrm{Re} \to \infty$. We attempt a solution of (2.17)–(2.19) an an expansion,[1]

$$u = u_0(x, y, t) + \cdots, \quad v = v_0(x, y, t) + \cdots, \quad p = p_0(x, y, t) + \cdots,$$
$$(3.3)$$

where the leading terms are governed by the Euler equations,

$$u_{0t} + u_0 u_{0x} + v_0 u_{0y} = -p_{0x}, \qquad (3.4)$$

$$v_{0t} + u_0 v_{0x} + v_0 v_{0y} = -p_{0y}, \qquad (3.5)$$

$$u_{0x} + v_{0y} = 0. \qquad (3.6)$$

The Euler equations are of lower order than the original Navier–Stokes system and require fewer boundary conditions. Keeping the no-penetration condition on the body surface, $v_0 = 0$ at $y = 0$ for $x \geq 0$, we have a simple solution,

$$u_0 = 1, \; v_0 = 0, \; p_0 = 0. \qquad (3.7)$$

3.2.   *The inner expansion — boundary layer*

The slip velocity in the outer solution (3.7) at $y = 0$ needs to be reduced to zero at the wall, as required by the no-slip conditions

(3.2). We therefore expect an inner solution to develop in a boundary layer — a slender region surrounding the wall (see Fig. 1(a)). An estimate for its thickness follows from the balance between advective and viscous terms in (2.17), $uu_x \sim \mathrm{Re}^{-1}u_{yy}$. With $u = O(1)$ and the x-scale also of $O(1)$ this gives $y = O(\mathrm{Re}^{-1/2})$. Also, from the continuity equation (2.19) we deduce that $v = O(\mathrm{Re}^{-1/2})$. We therefore define the scaled inner variable, $y = \mathrm{Re}^{-1/2}Y$, and attempt a solution in the form

$$u = U(x, Y, t) + \cdots, \quad v = \mathrm{Re}^{-\frac{1}{2}}V(x, Y, t) + \cdots, \quad p = P(x, Y, t) + \cdots.$$
(3.8)

On substitution into (2.17)–(2.19), we obtain the standard set of Prandtl boundary layer equations,

$$U_t + UU_x + VU_Y = -P_x + U_{YY},$$
(3.9)

$$P_Y = 0,$$
(3.10)

$$U_x + V_Y = 0.$$
(3.11)

As boundary conditions, we impose no-slip at the wall and the matching conditions with the outer solution,

$$U = 0, \; V = 0 \text{ at } Y = 0 \text{ for } x \geq 0,$$
(3.12)

$$U \to 1, \; P \to 0 \text{ as } Y \to \infty,$$
(3.13)

respectively. At this stage, we can already find the pressure function. From (3.10), $P = P(x, t)$, and the second matching condition in (3.13) shows that $P \equiv 0$.

Note that the uniform stream, $U = 1, V = 0$, is a valid solution inside the boundary layer upstream of the leading edge of the plate (i.e., the boundary layer does not propagate information upstream of the leading edge[b]). This leads to the initial condition,

$$U = 1 \text{ at } x = 0 \text{ for } Y > 0,$$
(3.14)

where we use symmetry and from now on confine ourselves to the upper half-plane, $Y \geq 0$.

The formulation (3.9)–(3.14) suggests a stationary solution. Also, with $P \equiv 0$, the boundary layer flow does not have a characteristic

---

[b]This feature is related to the parabolic nature of the boundary layer equations (as in the conventional diffusion equation).

length scale. This is a typical indication of a self-similar form of the solution which can be derived as follows. The continuity equation is satisfied by introducing the streamfunction, $\Psi(x, Y)$, such that $U = \partial\Psi/\partial Y, V = -\partial\Psi/\partial x$. Then taking $\Psi = x^\alpha f(\eta)$ with $\eta = Y/x^\beta$ and constant α and β, the momentum equation (3.9) takes the form

$$x^{2\alpha-2\beta-1} \left[(\alpha - \beta)(f')^2 - \alpha f f''\right] = x^{\alpha-3\beta} f'''. \qquad (3.15)$$

Since x and η are independent, we require $2\alpha - 2\beta - 1 = \alpha - 3\beta$. The boundary condition at the outer edge of the boundary layer (3.14) requires $\Psi \sim Y$ as $Y \to \infty$ or $x^\alpha f(\eta) \sim x^\beta \eta$ as $\eta \to \infty$. From this single condition two requirements emerge: first, $\alpha = \beta$ (and hence $\alpha = \beta = 1/2$ due to a relation between α and β derived earlier) and, second, $f(\eta) \sim \eta$ as $\eta \to \infty$. Perhaps surprisingly at first sight, the initial condition (3.14) gives no extra constraints on the solution. The wall conditions require that f and f' vanish at the wall at $\eta = 0$. Hence the similarity part of the streamfunction is determined from

$$f''' + \frac{1}{2}f f'' = 0; \quad f(0) = f'(0) = 0; \quad f(\eta) \sim \eta \text{ as } \eta \to \infty. \qquad (3.16)$$

The last formulation is known as the Blasius solution for a flat-plate boundary layer.

The graph of the velocity profile, $U = f'(\eta)$, is shown in Fig. 1(b). The boundary layer grows thicker downstream in accordance with the self-similar form, $Y = O(x^{1/2})$. In the near-wall part of the flow, $f(\eta) = \lambda_0 \eta^2/2 + \cdots$ as $\eta \to 0$, with $\lambda_0 = 0.332$ approximately. This means that the near-wall velocity in the boundary layer is given by

$$U(x, Y) = \lambda_0 x^{-\frac{1}{2}} Y + \cdots, \quad \text{as } Y \to 0, \qquad (3.17)$$

hence the wall shear (i.e., the tangential component of the stress vector) decreases downstream. It is also found that $f(\eta) = \eta - \delta^* + \cdots$ as $\eta \to \infty$, where δ^* is known as the displacement constant, and $\delta^* = 1.72$ approximately. The significance of a non-zero δ^* is evident when we consider the limit values of the transverse velocity $V(x, Y)$ at the outer edge of the boundary layer as $Y \to \infty$, namely $V = \delta^* x^{-1/2}/2 + \cdots$. This, from the second formula in (3.8), indicates that the outer expansion must contain terms of order $\text{Re}^{-1/2}$ which physically can be interpreted as the flow past a slender effective body of thickness $\mathrm{Re}^{-1/2}\delta^* x^{1/2}$.

3.3. *Classical boundary layers — hierarchy*

When the flat plate is replaced with a body of finite thickness the asymptotic procedure of this section remains the same, giving rise to a hierarchical sequence of approximations: compute the inviscid flow for the given body shape and find the slip velocity on the body surface, then solve for the boundary layer with the given velocity at the outer edge, return to the inviscid region and find the inviscid flow at the next order taking account of the displacement effect, and so on. This is the classical boundary layer approach which, despite being a necessary element of the flow description, nevertheless fails in a surprisingly large number of cases: at sharp trailing edges, near points of boundary layer separation on sufficiently thick bodies, at corner points in the body contour, at the leading edge of an airfoil at incidence, to name a few examples from external fluid dynamics. An alternative, interactive, boundary layer theory required in such situations is introduced in the following section.

4. Flow Past a Wall Roughness: The Triple-Deck

To understand how and why the classical boundary layer scheme may fail in a more complex flow, we introduce an irregularity which may vary in time (e.g., a hump or indentation) in the otherwise smooth wall shape and examine the response of the boundary layer flow to the wall roughness (see Fig. 2).

4.1. *Estimates*

The roughness is centered at $x = 1$ and has a characteristic non-dimensional length scale $L \ll 1$ (not to be confused with the dimensional length L used in Sec. 1) and height scale h. The roughness parameters can be more or less arbitrary, however not every choice leads to new or interesting flow responses. We aim to establish the

S. N. Timoshin

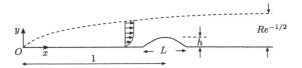

Fig. 2.   Flow past a hump on a flat plate.

length scale L sufficiently short to cause deviations from the standard boundary layer hierarchy and, at the same time, to choose the obstacle height h just large enough to provoke the first appearance of nonlinearity in the flow.

The flow ahead of the roughness is divided into two asymptotic layers: the inviscid current and the boundary layer of thickness $O(\text{Re}^{-1/2})$. In the undisturbed inviscid flow,

$$u = 1 + \cdots, \quad v = O(\text{Re}^{-\frac{1}{2}}), \quad p = O(\text{Re}^{-\frac{1}{2}}). \qquad (4.1)$$

Consider the inviscid part of the flow near the obstacle with characteristic dimensions $\Delta x \sim \Delta y \sim L$. As the flow traces the obstacle, a correction to the transverse velocity is induced, of magnitude $\Delta v \sim h/L$, assuming a shallow roughness with $h \ll L$. A similar order correction is also induced in the streamwise velocity, $\Delta u \sim h/L$, as follows from the continuity equation written as an order of magnitude balance, $\Delta u / \Delta x \sim \Delta v / \Delta y$. Then from the momentum balance, $u u_x + \cdots \sim p_x$, or $(1 + \cdots) \Delta u / \Delta x + \cdots \sim \Delta p / \Delta x$, we obtain an estimate for the pressure induced by the wall roughness, $\Delta p \sim h/L$. Moving on to the flow underneath the inviscid region, we can now evaluate the effect of the extra pressure Δp on the viscous sublayer. This is a very thin near-wall layer, of thickness δy say, where the incoming flow velocity is small, $u \sim y \text{Re}^{1/2}$. The balance of terms in the momentum equation, including viscous forces, requires

$$u u_x \sim p_x \sim \text{Re}^{-1} u_{yy}. \qquad (4.2)$$

The flow response is nonlinear when the velocity changes by some $\delta u \sim u$. With $\Delta p \sim h/L$, and taking $y \sim \delta y$, the terms in (4.2) are in balance provided that

$$u \sim L^{\frac{1}{3}}, \quad \Delta p \sim L^{\frac{2}{3}}, \quad y \sim L^{\frac{1}{3}} \text{Re}^{-\frac{1}{2}}, \quad h \sim L^{\frac{5}{3}}. \qquad (4.3)$$

Now we introduce into consideration one more effect. From the mass conservation in the viscous wall layer, $\delta u / L \sim \delta v / \delta y$, a change in the streamwise velocity induces a change in the transverse velocity component and the streamlines acquire an additional slope, ϕ say, estimated as

$$\phi \sim \frac{\delta v}{\delta u} \sim \frac{\delta y}{L} = \text{Re}^{-\frac{1}{2}} L^{-\frac{2}{3}}, \qquad (4.4)$$

where we have used the earlier estimate (4.3) for $\delta y \sim y$.

To recap, the wall roughness induces a pressure correction in the inviscid flow due to changes in the wall slope. The additional pressure acts on the viscous sublayer, changing the flow velocity and consequently changing once again the slope of the streamlines. If the two changes in the inclination of streamlines are comparable in magnitude, the flow in the viscous sublayer begins to interact with the flow in the inviscid region outside the boundary layer. Such a viscous-inviscid interaction takes place when $\phi \sim h/L$, i.e., when $L \sim \text{Re}^{-3/8}$. For the interaction regime, the flow quantities in the viscous sublayer are estimated as

$$y \sim \text{Re}^{-\frac{5}{8}}, \quad u \sim \delta u \sim \text{Re}^{-\frac{1}{8}}, \quad v \sim \text{Re}^{-\frac{3}{8}}, \quad \Delta p \sim \text{Re}^{-\frac{1}{4}}. \qquad (4.5)$$

With these estimates, the formal construction of the asymptotic expansions for the viscous-inviscid interaction is relatively straightforward.

4.2. *Outer region or upper deck*

It is convenient to introduce $\varepsilon = \text{Re}^{-1/8}$ as a small parameter. The flow has a triple-deck structure locally as shown in Fig. 3. Let $x = 1 + \varepsilon^3 X, t = \varepsilon^2 T$ with $(X, T) = O(1)$. In the upper deck, the normal coordinate is scaled according to $y = \varepsilon^3 y_1$, $y_1 = O(1)$, and the flow functions are perturbations to a uniform stream,

$$(u, v, p) = (1, 0, 0) + \varepsilon^2 \left(u_1(X, y_1, T), v_1(X, y_1, T), p_1(X, y_1, T) \right) + \cdots . \qquad (4.6)$$

From the Navier–Stokes equations (2.17)–(2.19), we have,

$$u_{1X} = -p_{1X}, \quad v_{1X} = -p_{1y_1}, \quad u_{1X} + v_{1y_1} = 0. \qquad (4.7)$$

Note that the flow in the upper deck is quasi-stationary.

From (4.7), one can easily derive Laplace's equation for p_1 or v_1, for instance, showing that the flow in the upper deck is potential.

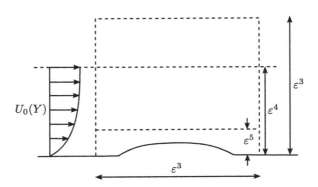

Fig. 3.   A schematic of the triple-deck region.

Alternatively, assuming that the effect of the obstacle is negligible far upstream, equations (4.7) can be manipulated into the Cauchy-Riemann conditions for a function, $F(z) = p_1 + iv_1$, analytic in the upper half plane of the complex variable $z = X + iy_1$. Then, from the Cauchy integral formula, we have,

$$F(z) = \frac{1}{2\pi i} \oint_C \frac{F(\zeta)}{\zeta - z} d\zeta, \qquad (4.8)$$

where z is a point in the upper half plane and the closed contour C consists of the x-axis and a semi-circle of an "infinitely" large radius in the upper half plane.

For the purposes of matching with the subsequent expansion in the middle deck (main part of the boundary layer), we need a relation between the flow quantities in the limit as $y_1 \to 0$. Such a relation, derived from (4.8), is given by the Sokhotski–Plemelji formula,

$$F(z) = \frac{1}{2}F(z) + \frac{1}{2\pi i}\int_{-\infty}^{\infty} \frac{F(\zeta)}{\zeta - z} d\zeta, \qquad (4.9)$$

where z is now a point on the boundary, i.e., $z = X$. The integration in (4.9) proceeds along the x-axis, and the principal value of the integral is taken. The real part of (4.9) gives a relation between the pressure function and the normal velocity at $y_1 = 0$,

$$p_1(X, 0, T) = \frac{1}{\pi} \int\limits_{-\infty}^{\infty} \frac{v_1(s, 0, T)}{s - X} ds. \tag{4.10}$$

A student of aerodynamics will probably recognise the last equation as a thin airfoil formula for a symmetric airfoil at a zero angle of attack.

4.3. *Main part of the boundary layer or middle deck*

Here, as in the undisturbed boundary layer arriving at the location of the wall roughness, we have $y = \varepsilon^4 Y$, $Y = O(1)$. The flow functions expand as

$$u = U_0(Y) + \varepsilon u_2(X, Y, T) + \cdots, \tag{4.11}$$

$$v = \varepsilon^2 v_2(X, Y, T) + \cdots, \quad p = \varepsilon^2 p_2(X, Y, T) + \cdots, \tag{4.12}$$

where $U_0(Y)$ is the streamwise velocity just ahead of the roughness. The governing equations for the first disturbance terms are

$$U_0 u_{2X} + v_2 U_0' = 0, \quad p_{2Y} = 0, \quad u_{2X} + v_{2Y} = 0 \tag{4.13}$$

with $U_0'(Y) = dU_0/dY$. For the velocity components, we can write the solution in the form,

$$u_2 = A(X, T) U_0'(Y), \quad v_2 = -A_X(X, T) U_0(Y) \tag{4.14}$$

with the so-called displacement function[c] $A(X, T)$ undetermined at this stage, whereas the pressure does not change across the middle deck, $p_2 = p_2(X, T)$. We can now perform matching between the upper and middle decks. For the pressure function, we have $p_2(X, T) = p_1(X, 0, T)$. For the vertical velocity, recall that $U_0(Y) \to 1$ as $Y \to \infty$, therefore $v_1(X, 0, T) = -A_X(X, T)$. This allows us to re-write (4.10) in the form

$$p_1(X, 0, T) = p_2(X, T) = -\frac{1}{\pi} \int\limits_{-\infty}^{\infty} \frac{\partial A(s, T)}{\partial s} \frac{ds}{s - X}. \tag{4.15}$$

---

[c]As follows from (4.11) and (4.14), the streamlines in the middle deck are displaced in the transverse direction by $-\varepsilon A(X, T)$, hence the terminology.

4.4.   *The viscous sublayer or lower deck*

Here we write $y = \varepsilon^5 y_3$, $y_3 = O(1)$. The flow functions expand as

$$u = \varepsilon u_3(X, y_3, T) + \cdots, \quad v = \varepsilon^3 v_3(X, y_3, T) + \cdots,$$
$$p = \varepsilon^2 p_3(X, y_3, T) + \cdots. \tag{4.16}$$

Substitution into the Navier–Stokes equations (2.17)–(2.19) gives the system of boundary layer equations,

$$u_{3T} + u_3 u_{3X} + v_3 u_{3y_3} = -p_{3X} + u_{3y_3 y_3}, \tag{4.17}$$

$$p_{3y_3} = 0, \tag{4.18}$$

$$u_{3X} + v_{3y_3} = 0. \tag{4.19}$$

Matching the pressure function with the solution in the main deck yields the pressure–displacement relation,

$$p_3(X, T) = -\frac{1}{\pi} \int\limits_{-\infty}^{\infty} \frac{\partial A(s, T)}{\partial s} \frac{ds}{s - X}. \tag{4.20}$$

Matching the streamwise velocity component, we have

$$u_3(X, y_3, T) = \lambda_0 y_3 + \lambda_0 A(X, T) + \cdots \text{ as } y_3 \to \infty, \tag{4.21}$$

using the property $U_0(Y) = \lambda_0 Y + \cdots$ as $Y \to 0$. Far upstream (on the triple-deck scale), the flow is expected to be unperturbed,

$$u_3 \to \lambda_0 y_3, \quad v_3 \to 0, \quad p_3 \to 0 \text{ as } X \to -\infty. \tag{4.22}$$

Suppose the wall obstacle is given by $y = \varepsilon^5 f((x-1)\varepsilon^{-3}, t\varepsilon^{-2})$, for some shape function f. In the scaled triple-deck variables, we then have the no-slip conditions in the form

$$u_3 = 0, \quad v_3 = \frac{\partial f}{\partial T} \quad \text{at } y_3 = f(X, T). \tag{4.23}$$

4.5.   *Prandtl transposition*

It is convenient to shift the transverse coordinate, y_3, to "flatten" the lower boundary of the solution domain. This is achieved by the change of variables known as the Prandtl transposition, $y_3 \longrightarrow y_3 +$

$f(X,T)$, $v_3 \longrightarrow v_3+f_T+u_3 f_X$, which leaves the system (4.17)–(4.20), (4.22), intact whereas (4.21) and (4.23) are replaced by,

$$u_3(X, y_3, T) = \lambda_0 y_3 + \lambda_0 A(X, T) + \lambda_0 f(X, T) + \cdots \text{ as } y_3 \to \infty, \tag{4.24}$$

$$u_3 = v_3 = 0 \quad \text{at } y_3 = 0, \tag{4.25}$$

respectively. The flow domain is now $y_3 \geq 0$, $-\infty < X < \infty$. The change in the transverse component of the velocity, v_3, has the effect of placing the observer into a frame of reference moving with the wall vertically (the term f_T) and also rotated locally by an angle $\arctan(f_X)$ to align the field of view with the wall roughness.

5. Examples of Triple-Deck Flows

On eliminating λ_0 by means of an affine transformation, and simplifying notation, the lower-deck equations (4.17)–(4.20), (4.22), (4.24), (4.25) from the previous section can be written as

$$u_t + uu_x + vu_y = -p_x + u_{yy}, \tag{5.1}$$

$$u_x + v_y = 0, \tag{5.2}$$

$$p = -\frac{1}{\pi} \int_{-\infty}^{\infty} \frac{\partial A(s, t)}{\partial s} \frac{ds}{s - x}, \tag{5.3}$$

$$x \to -\infty: \ u \to y, \ v \to 0, \ p \to 0, \tag{5.4}$$

$$y \to \infty: \ u = y + A(x, t) + f(x, t) + \cdots, \tag{5.5}$$

$$y = 0: \ u = 0, \ v = 0. \tag{5.6}$$

The triple-deck equations describe a rich variety of phenomena. We start with a few results which can be obtained by mostly analytical means.

5.1. *Linear flow regimes*

In the unperturbed flow with $f(x, t) = 0$, we have $u = y$, $v = p = A = 0$. Suppose we have a shallow roughness which performs time-periodic oscillations of frequency ω in the transverse direction,

$$f(x, t) = \hat{\delta}2 \cos(\omega t) f_a(x), \tag{5.7}$$

where the amplitude factor, $\hat{\delta}$, is small. The flow functions are expected to be perturbed by an amount of order $\hat{\delta}$,

$$u = y + \hat{\delta}u_1 + \cdots , \quad v = \hat{\delta}v_1 + \cdots , \quad p = \hat{\delta}p_1 + \cdots , \quad A = \hat{\delta}A_1 + \cdots \quad (5.8)$$

with the leading disturbance terms governed by the linear equations,

$$u_{1t} + yu_{1x} + v_1 = -p_{1x} + u_{1yy}, \tag{5.9}$$

$$u_{1x} + v_{1y} = 0, \tag{5.10}$$

$$p_1 = -\frac{1}{\pi} \int_{-\infty}^{\infty} \frac{\partial A_1(s,t)}{\partial s} \frac{ds}{s-x}, \tag{5.11}$$

$$x \to -\infty: \ u_1 \to 0, \ v_1 \to 0, \ p_1 \to 0, \tag{5.12}$$

$$y \to \infty: \ u_1 = A_1(x,t) + 2\cos(\omega t)f_a(x) + \cdots , \tag{5.13}$$

$$y = 0: \ u_1 = 0, \ v_1 = 0. \tag{5.14}$$

For a time-periodic solution,

$$[u_1, v_1, p_1, A_1] = e^{-i\omega t} [u_a(x,y), v_a(x,y), p_a(x), A_a(x)] + c.c., \tag{5.15}$$

using $c.c.$ for the complex conjugate. The problem is simplified further using Fourier transforms in x which we define, for a function $g(x)$, by the relations

$$\bar{g}(k) = \int_{-\infty}^{\infty} e^{-ikx}g(x)dx, \quad g(x) = \frac{1}{2\pi} \int_{-\infty}^{\infty} e^{ikx}\bar{g}(k)dk \tag{5.16}$$

for the transform and the inverse transform, respectively. This yields

$$(iky - i\omega)\bar{u}_a + \bar{v}_a = -ik\bar{p}_a + \frac{d^2\bar{u}_a}{dy^2}, \quad ik\bar{u}_a + \frac{d\bar{v}_a}{dy} = 0, \tag{5.17}$$

$$\bar{p}_a = |k|\bar{A}_a, \tag{5.18}$$

$$y \to \infty: \ \bar{u}_a = \bar{A}_a(k) + \bar{f}_a(k) + \cdots , \tag{5.19}$$

$$y = 0: \ \bar{u}_a = 0, \ \bar{v}_a = 0. \tag{5.20}$$

The transform of the Hilbert integral in (5.18) is obtained from the convolution theorem or, alternatively, by solving for the inviscid part of the triple-deck using Fourier transforms.

A step-by-step guide to the solution of problems similar to (5.17)–(5.20) appears elsewhere in this volume.[9] In brief, differentiating the momentum equation in (5.17) and using the continuity equation, the problem reduces to the Airy equation, $\tau''(\eta) - \eta\tau(\eta)=0$, for the shear, $\tau = d\bar{u}_a/dy$, with $\eta = y(ik)^{1/3} - i\omega(ik)^{-2/3}$. Choosing a suitable branch of $(ik)^{1/3}$, the solution satisfying the decay condition at infinity is found in terms of the Airy function, $\tau(\eta) = C_1 Ai(\eta)$, with a constant of integration, C_1, determined from the boundary conditions. From the pressure–displacement relation (5.18), the Fourier transform of the pressure function is then found in terms of the obstacle shape,

$$\bar{p}_a = \frac{|k|Ai'(-\zeta)\bar{f}_a(k)}{D(\omega,k)}, \tag{5.21}$$

where $\zeta = i\omega(ik)^{-2/3}$. The denominator in (5.21) is given by

$$D(\omega,k) = (ik)^{\frac{1}{3}}|k| \int_0^\infty Ai(s-\zeta)ds - Ai'(-\zeta). \tag{5.22}$$

Taking the inverse Fourier transform, we have the linearised pressure function,

$$p_1(x,t) = \frac{1}{2\pi} \int e^{i(kx-\omega t)} \frac{|k|Ai'(-\zeta)\bar{f}_a(k)}{D(\omega,k)} dk + c.c. \tag{5.23}$$

The last formula is interesting in two respects. First of all, the flow response to a moving wall roughness appears as a superposition of harmonic travelling waves of the form $p_1 \backsim \exp[i(kx - \omega t)]$ with the amplitudes determined by the wall roughness through the term $\bar{f}_a(k)$. Second, as is typical for evolution problems, the denominator in (5.23) contains the dispersion function of the system. Recall that by the dispersion relation we understand a functional relationship, $D(\omega,k) = 0$, between the wavenumber k and frequency ω of free waves which may exist in the flow without forcing. In our case, we can re-write (5.21) as

$$D(\omega,k)\bar{p}_a = |k|Ai'(-\zeta)\bar{f}_a. \tag{5.24}$$

Now, if the wall is undisturbed, $\bar{f}_a = 0$, then a non-trivial solution for the pressure exists if $D(\omega,k) = 0$.

5.2. *Properties of the dispersion relation*

The dispersion relation $D(\omega, k) = 0$ may be used to determine the frequency ω for a free travelling wave with a given wavenumber k (or wave length $2\pi/k$). With $D(\omega, k)$ given by (5.22), it is known that a countable set of free waves (wave modes) exists for any fixed $k \neq 0$, so that $\omega = \omega_n(k), n = 1, 2, 3, \ldots$. When k is real, the frequencies of free waves are complex valued. If we write $\omega = \omega_r + i\omega_i$ thus separating the real and imaginary parts, then, from the identity, $\exp[i(kx - \omega t)] = \exp[i(kx - \omega_r t)] \exp(\omega_i t)$, it follows that waves with $\omega_i < 0$ decay whereas waves with $\omega_i > 0$ grow which indicates instability. It turns out that in our case all wave modes except one are decaying. The diagram in Fig. 4 shows the loci of the first three roots in the complex plane ω for real $k > 0$. At small positive k the roots are clustered around the origin. As k increases, the second, third (and all subsequent roots) move deeper into the lower half plane and stay there for all k, hence showing decaying modes with $\omega_i < 0$. However, the first root crosses the real axis when $k = K_0 = 1.0005$ and $c = \omega/k = 2.2968$ approximately, which gives a neutral wave with $\omega = \Omega_0 = 2.298$. Waves with $k > K_0$ grow exponentially. We conclude that the triple-deck theory describes instability in the flow. Instability persists for higher k, in fact the growth rate ω_i for the first mode approaches a finite limit as $k \to \infty$.

5.3. *Oscillating wall*

In the flow due to an oscillating wall roughness, ω is real and the dispersion relation gives complex valued wavenumbers $k = k_r + ik_i$.

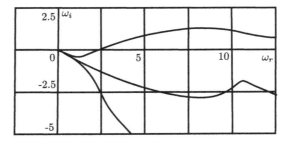

Fig. 4.   First three roots of the dispersion relation in the complex ω plane.

Waves with $k_i < 0$ grow in space as x increases since $\exp[i(kx - \omega t)] = \exp[i(k_r x - \omega t)] \exp(-k_i x)$ for a real ω.

It takes some analysis to prove that in the triple-deck flow waves cannot grow upstream, as $x \to -\infty$. The wave modes again form a countable set, $k = k_n(\omega), n = 1, 2, 3, \ldots$, for each real $\omega > 0$. Modes with $n > 1$ always decay downstream. The first mode decays downstream if $0 < \omega < \Omega_0$, it becomes neutral at $\omega = \Omega_0$ and grows downstream exponentially when $\omega > \Omega_0$. Perturbations with a sufficiently high frequency are said to display spatial instability (as opposed to temporal instability in the case of real k and complex ω).

From the discussion of the properties of the dispersion relation above, we can anticipate the flow response to a vibrating wall section in one of three forms. The roots of the dispersion relation are simultaneously the poles of the integrand in (5.23). In accordance with the Residue Theorem, a pole in the lower half-plane k gives an exponentially growing term as $x \to \infty$ downstream. It follows that a low frequency vibrator intiates a local disturbance field with the decay upstream and downstream as no poles are observed with $k_i < 0$. When the frequency of vibrations reaches the critical level, $\omega = \Omega_0$, the flow develops a neutral wave disturbance downstream. At higher frequencies, the response downstream is an exponentially amplified wave due to the pole with $k_i < 0$. This behaviour is illustrated in Fig. 5 adapted from the original study.[10]

It is important to understand the meaning of the triple-deck instability in the more general context of the flow at finite Reynolds numbers. A traditional and straightforward (although not unquestionable) approach to boundary layer stability calculations involves a linearisation of the Navier–Stokes equations about the Blasius velocity profile assuming that the undisturbed flow is independent of x. This leads to an eigenvalue problem for the disturbance wavenumber/frequency pair with the Reynolds number acting as a parameter (the Orr–Sommerfeld theory).[11] The result of such calculations is presented as a neutral curve in the Reynolds number — frequency plane shown with a solid line in Fig. 6. In the parameter region inside the neutral curve disturbances grow, the region outside corresponds to decaying disturbances. Note that the Orr–Sommerfeld neutral curve

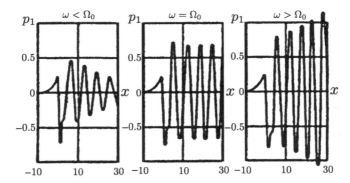

Fig. 5.   Pressure past the oscillator for subcritical ($\omega < \Omega_0$), neutral ($\omega = \Omega_0$) and supercritical ($\omega > \Omega_0$) frequencies. The oscillating section of the wall has a triangular shape located in $0 \leq x \leq 2$.

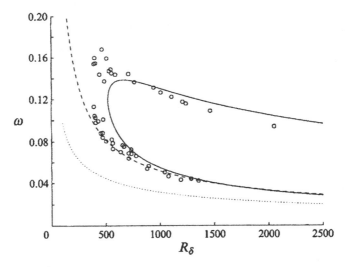

Fig. 6.   Blasius boundary layer stability. Solid: Orr–Sommerfeld solutions; dots and dashes: first- and second-order triple-deck; dots: experiments.

uses the Reynolds number R_δ based on the boundary layer thickness as the length scale rather than our global Reynolds number.

The triple-deck theory captures the flow properties at asymptotically large Reynolds numbers. The neutral wave parameters computed from the triple-deck equations are shown in Fig. 6 with dots (the dashed curve is the result of a higher order triple-deck computation).[12] Clearly, the triple-deck model reproduces the stability characteristics near the lower branch of the Orr–Sommerfeld neutral curve. The circles in the same figure represent some experimental data, again in general agreement with theoretical results. We should emphasise, however, that the triple-deck scheme does not capture the upper branch of the stability curve where an alternative asymptotic theory is needed (for further discussion see Ref. 12 and references therein).

5.4. *Boundary layer separation*

In nonlinear flow regimes, the triple-deck equations describe boundary layer separation and associated phenomena (non-uniqueness, secondary instability, etc.). To begin with, consider the flow near a corner point in the wall geometry. An illustration of the numerical solution of the triple-deck equations for a steady flow with $f = 0$ for $x < 0$ and $f = \alpha x$ for $x > 0$ is shown in Fig. 7.[4] The value of the scaled angle parameter, $\alpha = 7$, in this example is quite extreme — it is expected that the flow will show a local separation region in the corner for a sufficiently large α, however, here we observe also a secondary separation. It was suggested that the solution does not exist above a certain critical value of α, possibly due to a singularity inside the separation region.[13] Moreover, solutions for negative values of α are not unique[d] and also exist in a finite range of the angle parameter. It would be

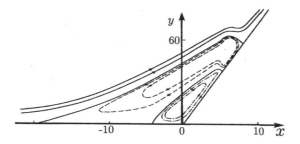

Fig. 7.   Flow past a corner, the scaled wall shape is $y = 7x$.

---

[d]Here, the asymptotic theory gives a direct validation of non-uniqueness of the Navier–Stokes solutions for a sufficiently large Reynolds number.

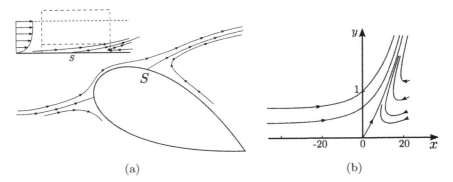

Fig. 8.   Laminar separation (a) Separation on the global and triple-deck scales, (b) triple-deck solution in the viscous sublayer.

certainly interesting to find out whether the non-existence of a steady triple-deck state signifies an alternative steady flow (with a different asymptotic structure at large Reynolds numbers) or perhaps a necessary transition to a time-dependent (maybe even turbulent) flow. At the same time, it is known that the corner flow displays absolute (as opposed to convective) instability for large α so it may be quite difficult to study the flow evolution within the time-dependent triple-deck framework.[14]

Another important phenomenon described by the triple-deck theory is the boundary layer separation from a blunt body as illustrated in Fig. 8(a). The asymptotic flow structure in this flow is rather complicated. In the inviscid part, the streamline leaving the body surface at the separation point marks the boundary between the incoming potential stream and the region of a nearly stagnant fluid downstream. The potential flow is now chosen from a class of discontinuous solutions of the Euler equations with a free streamline. Such solutions are not unique and the "correct" solution is selected by local conditions in the viscous flow near the separation point. The local viscous flow is governed by the triple-deck equations with a solution in the viscous sublayer illustrated in Fig. 8(b). The flow separation here appears as a self-induced, spontaneous, process (triggered, of course, by the global flow geometry).[4]

6. Other Multi-Deck Interactive Flows

6.1. *Near-wall viscous jets: Double-deck*

If fluid is injected along a solid boundary into a quiescent environment, a near-wall jet is formed, of thickness $O(\text{Re}^{-1/2})$ in non-dimensional variables with a typical jet profile as shown in Fig. 9. In the case of a flat wall, for example, the jet emerging from a narrow slit will have the self-similar Glauert velocity profile, $u = x^{-1/2} F(y\text{Re}^{1/2}x^{-3/4})$. We are not concerned with the jet formation and evolution and focus instead on a double-deck interactive flow initiated by a local wall roughness. The main new feature here, compared with the standard triple-deck flow, is the pressure variation across the main part of the jet. In the non-dimensional Navier–Stokes equations (2.17)–(2.19), we introduce local variables, $x = 1 + \text{Re}^{-3/7}X$, $t = \text{Re}^{-2/7}T$, assuming a wall roughness positioned at $x = 1$. In the **main part** of the jet, $y = \text{Re}^{-1/2}Y$, the flow functions expand as,

$$u = U_0(Y) + \text{Re}^{-\frac{1}{7}}u_1(X, Y, T) + \cdots, \tag{6.1}$$

$$v = \text{Re}^{-\frac{2}{7}}v_1(X, Y, T) + \cdots, \quad p = \text{Re}^{-\frac{2}{7}}p_1(X, Y, T) + \cdots. \tag{6.2}$$

The perturbed velocity components, in the form $u_1 = A(X, T) U_0'(Y)$, $v_1 = -A_X(X, T)U_0(Y)$, follow readily from the streamwise momentum and continuity equations. There is a clear similarity with

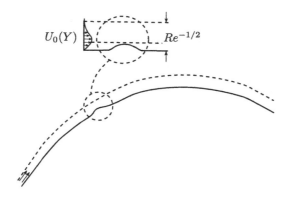

Fig. 9. Wall jet near a local obstacle.

the perturbed flow field in the main deck of the triple-deck flow (4.14). The transverse momentum balance, however, is significantly different giving

$$U_0 v_{1X} = -p_{1Y} \qquad (6.3)$$

with the solution for the pressure in the general form,

$$p_1 = -A_{XX}(X,T) \left[\kappa - \int_0^Y U_0^2(s)ds \right] + C(X,T),$$

$$\text{where } \kappa = \int_0^\infty U_0^2(s)ds. \qquad (6.4)$$

From the decay condition at the outer edge of the boundary layer, $C(X,T) = 0$. Then as $Y \to 0$ we have $p_1 \to -\kappa A_{XX}(X,T)$. Also, $U_0 = \lambda_0 Y + \cdots$, therefore $u_1 = \lambda_0 A(X,T) + \cdots$ as $Y \to 0$.

In the **viscous sublayer**, we have $y = \mathrm{Re}^{-9/14} y_2$, $y_2 = O(1)$, with the expansions

$$u = \mathrm{Re}^{-\frac{1}{7}} u_2(X, y_2, T) + \cdots, \quad v = \mathrm{Re}^{-\frac{5}{14}} v_2(X, y_2, T) + \cdots, \quad (6.5)$$

$$p = \mathrm{Re}^{-\frac{2}{7}} p_2(X, y_2, T) + \cdots. \qquad (6.6)$$

From the Navier–Stokes equations, it follows that $p_2 = p_2(X,T)$ and also

$$u_{2T} + u_2 u_{2X} + v_2 u_{2y_2} = -p_{2X} + u_{2y_2 y_2}, \quad u_{2X} + v_{2y_2} = 0. \qquad (6.7)$$

Matching with the flow in the main part of the jet and including the conditions in the incoming stream, we have

$$p_2 = -\kappa A_{XX}, \quad \text{and} \quad u_2 = \lambda_0 y_2 + \lambda_0 A(X,T) + \cdots \quad \text{as } y_2 \to \infty, \qquad (6.8)$$

$$u_2 \to \lambda_0 y_2 \quad \text{as } X \to -\infty. \qquad (6.9)$$

The no-slip conditions can be written as,

$$u_2 = v_2 = 0 \quad \text{at } y_2 = f(X), \qquad (6.10)$$

where the local irregularity is represented by $f(X)$. With the constants λ_0, κ eliminated by means of a suitable change of variables (see exercises below), the viscous-inviscid interaction problem takes the form,

$$\tilde{u}_{\tilde{t}} + \tilde{u}\tilde{u}_{\tilde{x}} + \tilde{v}\tilde{u}_{\tilde{y}} = -\tilde{p}_{\tilde{x}}(\tilde{x},\tilde{t}) + \tilde{u}_{\tilde{y}\tilde{y}}, \quad \tilde{u}_{\tilde{x}} + \tilde{v}_{\tilde{y}} = 0, \quad \tilde{p} = -\tilde{A}_{\tilde{x}\tilde{x}}(\tilde{x},\tilde{t}), \tag{6.11}$$

$$\tilde{u} = \tilde{y} + \tilde{A}(\tilde{x},\tilde{t}) + \cdots \quad \text{as } \tilde{y} \to \infty, \qquad \tilde{u} \to \tilde{y} \quad \text{as } \tilde{x} \to -\infty, \tag{6.12}$$

$$\tilde{u} = \tilde{v} = 0 \quad \text{at } \tilde{y} = \tilde{f}. \tag{6.13}$$

The tilde is used here to label the scaled variables. Mathematically, this formulation is closely related to the interactive flow problem for an asymmetric channel.[9] The most significant difference in the wall jet interaction compared to the standard triple-deck (5.1)–(5.6) is in the pressure–displacement relation. The principal-value integral in the triple-deck, stemming from the potential flow in the upper deck, introduces a degree of ellipticity in the problem formulation (in particular, local perturbations tend to lead to algebraically decaying disturbance fields upstream and downstream). The wall jet interaction formula, $\tilde{p} = -\tilde{A}_{\tilde{x}\tilde{x}}$, is local, at least in appearance. This does not mean however that the interactive flow replicates fully the boundary layer flow with a specified pressure (as in the Blasius solution). The essential new feature is the ability of the interactive boundary layer to propagate a disturbance upstream of the disturbance source, as we show now.

Upstream influence. Assuming a shallow roughness, $\tilde{f} = \delta f_l(\tilde{x})$, with $\delta \ll 1$, we linearise the equations of motion taking

$$\{\tilde{u},\tilde{v},\tilde{p},\tilde{A}\} = \{\tilde{y},0,0,0\} + \delta\{u_l(\tilde{x},\tilde{y}), v_l(\tilde{x},\tilde{y}), p_l(\tilde{x}), A_l(\tilde{x})\} + O(\delta^2), \tag{6.14}$$

so that the formulation (6.11)–(6.13) reduces to,

$$\tilde{y}u_{l\tilde{x}} + v_l = A_l'''(\tilde{x}) + u_{l\tilde{y}\tilde{y}}, \quad u_{l\tilde{x}} + v_{l\tilde{y}} = 0, \tag{6.15}$$

$$u_l \to A_l(\tilde{x}) \quad \text{as } \tilde{y} \to \infty, \qquad u_l \to 0 \quad \text{as } \tilde{x} \to -\infty, \tag{6.16}$$

$$u_l = -f_l(\tilde{x}), \quad v_l = 0 \quad \text{at } \tilde{y} = 0. \tag{6.17}$$

The problem (6.15)–(6.17) is solved using Fourier transforms, as in the previous section. In particular, for the pressure function we have,

$$p_l(\tilde{x}) = -\frac{a}{2\pi} \int_{-\infty}^{\infty} \frac{(ik)^2 \exp(ik\tilde{x}) \bar{f}_l(k)}{(ik)^{\frac{7}{3}} - a} dk, \qquad (6.18)$$

where $a = 3|Ai'(0)|$ and $\bar{f}_l(k)$ is the Fourier transform of the wall roughness shape. We aim to evaluate the response of the flow upstream of the obstacle, $X < 0$, which is done by closing the contour of integration in (6.18) in the lower half plane and using the Residue Theorem, with the pole at $k = -ia^{3/7}$. The result is

$$p_l(\tilde{x}) = \frac{3}{7} a^{\frac{9}{7}} \bar{f}_l(-ia^{\frac{3}{7}}) \exp(a^{\frac{3}{7}} \tilde{x}) \text{ for } \tilde{x} < 0. \qquad (6.19)$$

We conclude that the flow "feels" the presence of an obstruction at distances proportional to the length scale of the interaction region.

Self-induced separation. Suppose there are no obstacles but the flow in the jet viscous sublayer deviates from the unperturbed base state following the eigenfunction (6.19) far upstream. This looks a little like a bifurcation phenomenon, with the base flow as the undisturbed state and two families of neighbouring solutions branching from the base state in accordance with the sign of the (arbitrary) coefficient in front of the eigenfunction. The solution with falling pressure was found to terminate in a finite-distance singularity with unbounded acceleration of the fluid particles close to the wall. The solution branch with rising pressure leads to a self-induced separation with the velocity profiles shown in Fig. 10.[15] Note that the wall shear ahead of the interaction region is $1/2$ in the figure.

6.2. *Supersonic self-induced separation*

Viscous-inviscid interaction in a supersonic flow was described by Stewartson and Williams and Neiland around 1970. It is governed by Eqs. (6.11)–(6.13) or Eqs. (5.1)–(5.6) but with the pressure-displacement relation,

$$\tilde{p} = -\tilde{A}_{\tilde{x}}(\tilde{x}, \tilde{t}). \qquad (6.20)$$

This is another local interaction which displays upstream influence and self-induced separation. The supersonic triple-deck does not

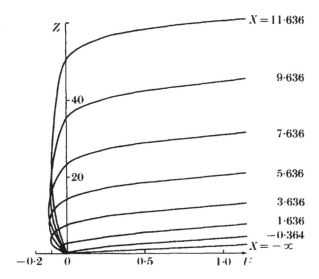

Fig. 10.   Velocity profiles in the separation of a wall jet.

describe instability of the flow at a large Reynolds number. The technique of linearising the sublayer equations can be used to prove the existence of upstream influence in the flow (see an exercise at the end of this section) and hence deduce the possibility of self-induced processes in the boundary layer. Self-induced separation is illustrated in Fig. 11.[16]

6.3.   *Condensed flow*

In many applications, an interactive flow is characterised by the absence of displacement effect in the viscous sublayer. We shall introduce the notion of such a condensed flow as a short-scale limit of the triple-deck (5.1)–(5.6). Let $x \ll 1$. The balance of terms in the momentum equation (5.1) is maintained if $u \sim y \sim x^{1/3}$, $p \sim x^{2/3}$. If we assume $A \sim x^{1/3}$ to keep the displacement term in (5.5) then we would have $p \sim x^{-2/3}$ according to (5.3), in contradiction with the earlier estimate. The contradiction is avoided if (5.3) is replaced with

$$A(x, t) = 0. \qquad (6.21)$$

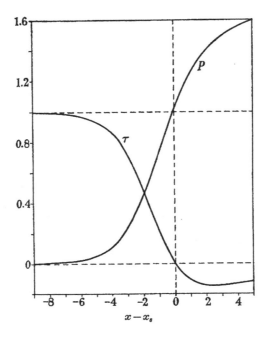

Fig. 11.   Pressure and wall shear in the supersonic separation.

The pressure in the boundary layer remains unknown therefore we have a non-classical, interactive, boundary layer behaviour. It is easy to verify that the 2D condensed flow does not support upstream influence. However, upstream influence is present in three dimensions. These properties are left as an exercise although a comprehensive analysis with applications can be found in Ref. 9.

7. Conclusions

Asymptotic solutions of the Navier–Stokes equations have a clear multi-deck structure and lead to a considerable simplification of the original problem often allowing closed-form analytical or inexpensive numerical solutions. There is a penalty, however. Due to the nature of the limit processes involved, the accuracy of the solution is not known in advance and needs to be tested either against an alternative (most likely, numerical) solution of the full problem or against experimental observations. This needs to be taken into account applying the methods of analysis summarised in this chapter in new areas.

Exercises (the first digit refers to the section number)

Exercise 2.1 Verify the formula (2.3) for a one-dimension (1D) motion with $\delta V = \delta x$.

Exercise 2.2 Verify the product rule,

$$\frac{D\,(\mathbf{a} \cdot \mathbf{b})}{Dt} = \frac{D\mathbf{a}}{Dt} \cdot \mathbf{b} + \mathbf{a} \cdot \frac{D\mathbf{b}}{Dt}.$$

Exercise 2.3 The flow is two-dimensional (2D), bounded by a flat plate at $y = 0$. Write down the components of the stress tensor on the boundary in terms of the velocity and pressure.

Exercise 3.1 Falkner–Skan boundary layer. The boundary layer equations (3.9–3.11) are solved subject to the no-slip conditions (3.12) and with the outer-edge condition $U \to U_0 x^m$ as $Y \to \infty$. Find the pressure in the boundary layer and suggest a similarity form for the solution. Do such boundary layers have a physical meaning?

Exercise 3.2 The flat plate in Sec. 3 has length 1. Find the solution in the classical boundary layer immediately downstream of the trailing edge, $0 \le x-1 \ll 1$? What is the solution in the boundary layer far downstream? A second plate is inserted in the wake parallel to the first plate. What can be said about the boundary layer on the second plate?

Exercise 4.1 Show that the undisturbed shear coefficient, λ_0, can be eliminated from the sublayer Eqs. (4.17)–(4.23) by an affine transformation.

Exercise 5.1 The solution in Sec. 5.3 grows exponentially downstream at super-critical frequencies, $\omega > \Omega_0$. How can it be obtained using Fourier transforms?

Exercise 5.2 Derive the formula (5.18) for the Fourier transform of the pressure–displacement relation by solving for the flow in the upper deck of the triple-deck, Eqs. (4.7), using Fourier transforms and appropriate boundary conditions.

Exercise 5.3 Verify that the dispersion relation, $D(\Omega, k) = 0$, with D given in (5.22) follows from the linearised triple-deck equations directly, i.e., not assuming a wall roughness as in Sec. 5.1.

Exercise 5.4 Near the separation point in Fig. 7 or Fig. 8(b), with $x = x_s$ say, the streamwise velocity is of the form, $u = ay^2 - b(x - x_s)y + \cdots$, with positive constants a and b. Downstream of x_s, the flow field admits exponentially small eigenfunctions of the form,

$$u = ay^2 - b(x - x_s)y + \cdots + C \exp\left(-\frac{\alpha}{x^\rho}\right) f\left(\frac{y}{x}\right) + \cdots,$$

where α, ρ are positive and C is arbitrary. Using (5.1)–(5.6), determine the value of ρ.

Exercise 6.1 Find an affine transformation, $x_2 = a_1 \tilde{x}$, $y_2 = a_2 \tilde{y}$, $f = a_2 \tilde{f}$, $u_2 = a_3 \tilde{u}$, $v_2 = a_4 \tilde{v}$, $A = a_5 \tilde{A}$, $p_2 = a_6 \tilde{p}$, $T = a_7 \tilde{T}$ with constants a_{1-7} such that the parameters λ_0 and κ are scaled out from (6.7)–(6.10).

Exercise 6.2 Derive a linearised solution for the supersonic flow with the pressure–displacement relation (6.20) past a shallow obstacle and hence show the existence of upstream influence.

Exercise 6.3 Consider interactive flows with the pressure–displacement relations $p = A$ and $p = -A$. Which of these two interactions supports upstream influence in a 2D case?

Solutions

Solution 2.1 Let $\delta V = x_2(t) - x_1(t)$ then $D(\delta x)/Dt = u_2 - u_1 = \delta u$, where $u_{1,2} = Dx_{1,2}/Dt$ since $x_{1,2}$ are coordinates of fluid particles. Hence $(\delta x)^{-1} D(\delta x)/Dt \to \partial u/\partial x$ in the limit $\delta x \to 0$.

Solution 2.2 D/Dt is a linear differential operator, hence the result.

Solution 2.3 $\sigma_{xx} = -p$, $\sigma_{yy} = -p$, $\sigma_{xy} = \sigma_{yx} = \mu u_y$, keeping in mind $u = v = 0$ at $y = 0$ (and the continuity equation).

Solution 3.1 Use $P_x = -\lim_{Y \to \infty} UU_x = -U_0^2 m x^{2m-1}$ and integrate. Method is similar to the Blasius solution, in particular $U = x^m f'(Y x^{(m-1)/2})$. The similarity equation for f is, of course, different from Blasius. This applies to flow past a wedge.

Solution 3.2 In the near wake, the flow near the symmetry line has the streamfunction $\Psi = (x - 1)^{2/3} g(Y(x - 1)^{-1/3})$ with $g''' +$

$2/3gg'' - 1/3(g')^2 = 0$, $g(0) = g''(0) = 0$, $g(\eta) = O(\eta^2)$ as $\eta \to \infty$. In the far wake, $U = 1 + Cx^{-1/2}\exp(-Y^2/4x) + \cdots$, with constant C, as obtained by linearising the boundary layer equations about the uniform stream. A second plate would be in a rotational incoming stream hence a non-uniform starting velocity profile. But if the plates are offset in the transverse direction by a distance $\gg \mathrm{Re}^{-1/2}$ then the effect of the first plate can be neglected.

Solution 4.1 The method is similar to Exercise 6.1.

Solution 5.1 The contour of integration needs to dip below the real axis to walk around the pole in the lower half plane k.

Solution 5.2 Introduce the potential, ϕ, such that $u_1 = \phi_X$, $v_1 = \phi_{y_1}$, hence $\phi_{XX} + \phi_{y_1 y_1} = 0$. Applying the Fourier transform (5.16) in X, the transformed potential satisfies $-k^2\bar{\phi} + \bar{\phi}_{y_1 y_1} = 0$, with the general solution $\bar{\phi} = C_1\exp(-|k|y_1) + C_2\exp(|k|y_1)$. Boundedness in the far field requires $C_2 = 0$. Then at $y_1 = 0$, we have $\bar{p}_1 = -\bar{u}_1 = -ikC_1$, $\bar{v}_1 = -|k|C_1$. Next, using $v_1|_{y_1=0} = -\partial A/\partial X$, or $\bar{v}_1|_{y_1=0} = -ik\bar{A}$, derive $\bar{p}_1|_{y_1=0} = |k|\bar{A}$. Finally, note that the pressure function does not change across the middle deck.

Solution 5.3 In (5.9)–(5.14), write $u_1 = \bar{u}_a\exp[i(kx - \omega t)] + c.c.$, with similar expressions for the other flow functions. The formulation reduces to (5.17)–(5.20) with $\bar{f}_a = 0$. Hence the result.

Solution 5.4 From the balance of convective and viscous terms, $\rho = 3$. For a complete analysis, see Ref. 17.

Solution 6.1 From (6.8), $a_5 = a_2$ and $a_3 = \lambda_0 a_2$. In the continuity equation, require $a_4 = \lambda_0 a_2^2/a_1$. Then for the time-derivative term in (6.7), require $a_7 = a_1/(\lambda_0 a_2)$ and the balance between advection and pressure gradient terms gives $a_6 = (\lambda_0 a_2)^2$ and the balance with the viscous term leads to $a_1 = \lambda_0 a_2^3$. The relation unused so far is the first formula in (6.8), where we find $\lambda_0^2 a_1^2 a_2 = \kappa$. Solving the last two equations for a_1, a_2 we find $a_1 = \kappa^{3/7}\lambda_0^{-5/7}$, $a_2 = \kappa^{1/7}\lambda_0^{-4/7}$, and the rest of the constants now follow.

Solution 6.2, 6.3 For a steady flow, consider the linearised formulation,

$$yu_x + v = -p'(x) + u_{yy}, \qquad u_x + v_y = 0, \tag{7.1}$$
$$y \to \infty: \quad u \to A(x) + F(x); \qquad y = 0: \quad u = v = 0. \tag{7.2}$$

In Fourier transforms, $iky\bar{u} + \bar{v} = -ik\bar{p} + \bar{u}_{yy}$, hence $iky\bar{u}_y = \bar{u}_{yyy}$, with the solution, $\bar{u}_y = C(k)Ai(y(ik)^{1/3})$. From the wall conditions, $C(k)(ik)^{1/3}Ai'(0) = ik\bar{p}$. The condition at infinity yields $C(k) = 3(ik)^{1/3}(\bar{A} + \bar{F})$, using the property $\int_0^\infty Ai(s)ds = 1/3$. Eliminating $C(k)$ we have, $(ik)^{1/3}\bar{p} + a\bar{A} = -a\bar{F}$. Here, $a = -3Ai'(0) > 0$ as it is known that $Ai'(0) < 0$. For the supersonic interaction, $\bar{p} = -ik\bar{A}$ hence $\bar{A} = -a\bar{F}[a - (ik)^{4/3}]^{-1}$, with a simple pole in the lower half-plane at $k = -ia^{3/4}$. This gives upstream influence. For the interactions given by $\bar{p} = \pm\bar{A}$, we have $\bar{A} = -a\bar{F}[a \pm (ik)^{1/3}]^{-1}$, therefore upstream influence in the case $p = -A$.

References

1. H. J. Wilson, Practical analytical methods for partial differential equations, in *Appl. Math.: Fluid and Solid Mechanics*, ICP (2015).
2. F. T. Smith, On the high Reynolds number theory of laminar flows. *IMA J. Appl. Math.* **28(3)**, 207–281 (1982).
3. F. T. Smith and A. P. Rothmayer, Incompressible triple-deck theory, in Johnson, R. W. (ed.) *CRC Handbook of Fluid Dynamics*, Dover Publications, Inc, New York, 1998.
4. V. V. Sychev, A. I. Ruban, Vic. V. Sychev and L. G. Korolev, *Asymptotic Theory of Separated Flows*, Cambridge University Press, Cambridge, 1998.
5. I. J. Sobey, *Introduction to Interactive Boundary Layer Theory*, Oxford University Press, Oxford, 2000.
6. G. K. Batchelor, *An Introduction to Fluid Dynamics*, Cambridge University Press, Cambridge, 2000.
7. L. D. Landau and E. M. Lifshitz, *Fluid Mechanics*. Second Edition: Vol 6 (Course of Theoretical Physics) Butterworth-Heinemann, 1987, pp. 554.
8. H. Schlichting and K. Gersten, *Boundary-Layer Theory*, Springer, 2000.
9. F.T. Smith, Internal fluid dynamics, in *Applied Mathematics: Fluid and Solid Mechanics*. ICP, 2015.
10. E. D. Terent'ev, The linear problem of a vibrator performing harmonic oscillations at supercritical frequencies in a subsonic boundary layer. *Prikl. Matem. Mekh. USSR*, **48(2)**, 184–191 (1984).
11. P. G. Drazin and W. H. Reid, *Hydrodynamic Stability*, Cambridge University Press, Cambridge, 2004.
12. J. J. Healey, On the neutral curve of the flat-plate boundary layer: comparison between experiment, Orr-Sommerfeld theory and asymptotic theory. *J. Fluid Mech.* **288**, 59–73 (1995).
13. F. T. Smith, A reversed-flow singularity in interacting boundary layers. *Proc. R. Soc. Lond. A* **420**, 21–52 (1988).

14. R. P. Logue, J. S. B. Gajjar and A. I. Ruban, Global stability of separated flows: subsonic flow past corners. *Theoret. Comput. Fluid Dyn.* **25**, 119–128 (2011).
15. F. T. Smith and P. W. Duck, Separation of jets or thermal boundary layers from a wall. *Q. J. Mech. Appl. Math.* **30(2)**, 143–156 (1977).
16. K. Stewartson and P. G. Williams, Self-induced separation. *Proc. R. Soc. Lond.* A **312**, 181–206 (1969).
17. F. T. Smith, Concerning upstream influence in separating boundary-layers and downstream influence in channel flow. *Q. J. Mech. Appl. Math.* **37**, 389–399 (1984).

Chapter 3

Time-mean Turbulent Shear Flows: Classical Modelling — Asymptotic Analysis — New Perspectives

Bernhard Scheichl[*]

[†]*Vienna University of Technology,*
Institute of Fluid Mechanics and Heat Transfer,
Tower BA/E322, Getreidemarkt 9, 1040 Vienna, Austria,
bernhard.scheichl@tuwien.ac.at

This contribution gives an introduction to the description of time-averaged single-phase turbulent flows where the largeness of the typical Reynolds numbers at play is their essential characteristic. Hence, the full Navier–Stokes equations form the starting point, and the viewpoint is the most rigorous asymptotic one. Emphasis is placed on the aspects of modelling the unclosed terms in the accordingly Reynolds-averaged equations when governing slender shear flows. Such represent the natural manifestation of turbulence as triggered internally in laminar shear layers by the no-slip condition to be satisfied at rigid walls rather than by free-stream turbulence, neglected here. Given the inherent closure problem associated with the separation and interaction of the variety of spatial/temporal scales involved, this focus allows for a surprisingly deep understanding of turbulent flows resorting to formal asymptotic techniques under the premise of a minimum of reliable assumptions. These are motivated by physical intuition and/or based on classical findings of the statistical theory of locally isotropic turbulence. Intrinsic differences to the analysis of related problems dealing with laminar high-Reynolds-number flows are highlighted. Finally, the crucial aspects of numerical simulation of turbulent flows are considered for the staggered levels of filtering, ranging from a most complete resolution to full averaging.

---

[*]Docent for fluid mechanics. Supplemental material can be delivered on request.
[†]Also: AC^{2}T research GmbH, Viktor-Kaplan-Straße 2/C, 2700 Wiener Neustadt, Austria.

1. Introduction

Without doubt, turbulence is one of the most fascinating and likewise vital fields of modern fluid dynamics, if not still going to be the most challenging of all in classical physics. Needless to say, gaining a deep understanding of turbulent flows is also of tremendous importance because of their omnipresence in various engineering applications.

1.1. *Prerequisites, objectives, scope, further reading*

The chapter is motivated by the exciting challenge addressed above and thus intends to highlight *some selected* intriguing phenomena associated with slender shear flows and topics of the current research from a most rigorous (asymptotic) viewpoint, some in a previously unappreciated manner. To this end, we assume a suitably (globally) defined Reynolds number, Re, to take on arbitrarily large values. Hence, the goal is definitely *not* to provide an overview on the entire subject (which is a "mission impossible" at all) but to encourage the interested reader to delve deeper into specific topics. He/she is expected to be familiar with fundamentals of fluid mechanics (including turbulence), dimensional analysis and perturbation methods.

We complete these introductory aspects by pointing to some classical and modern textbooks on turbulence as definite references. (This list must remain incomplete given the fast development of the field; to a certain extent, it naturally factors in the author's personal interests and approach and his integration in the Continental evolution of modern fluid dynamics and what is acknowledged as "Viennese school of asymptotics".)

Reference 1 provides a definitely pioneering and classical overview on the essence of shear-flow turbulence in a Newtonian fluid, appealing newcomers. Their specific peculiarities in terms of scaling laws are considered in Refs. 2 and 3, the latter taking up an asymptotic viewpoint and also giving an comprehensive survey on the commonly adopted turbulence models. More general and recent approaches are provided by Refs. 4–6; modelling aspects are dealt with extensively in Refs. 7–9. The biographic (non-asymptotic) view in Ref. 10, albeit probably not attracting the novice, deserves attention as well. Some further exposures and reviews of more specific topics are cited in due place.

1.2. *Preliminaries*

The enormous difficulties we are facing in the most rigorous treatment of turbulent flows — theoretically and/or numerically — lie in a specific property of the underlying Navier–Stokes equations (NSEs) in suitable non-dimensional form, thus entered by Re as the essential parameter: for unsteady flow, the subtle interplay between the nonlinear convective and the viscous term, at first considered as of $O(\nu_r)$ with the non-dimensional reference viscosity $\nu_r := \mathrm{Re}^{-1}$, entails a cascade of temporal/spatial scales involved: these range from the largest ones fixing Re and here of $O(1)$, subsequently referred to as global ones, to, on the other extreme, the smallest ones responsible for the conversion of the work exerted by internal viscous forces into internal energy, the so-called viscous dissipation. The well-known little mnemonic rhyme by the renowned physicist and meteorologist L. F. Richardson from ca. 1930 condenses nicely this process associated with separation of scales:

> *Big eddies have little eddies/and little eddies have smaller eddies/that feed on their vorticity/and so on to viscosity.*

As a crucial observation, this not only becomes the more pronounced the larger Re is — which justifies an asymptotic approach — but cannot be inferred *a piorily* from inspection of the NSE when subjected to decent initial/boundary conditions (ICs/BCs). In other words, it represents an inherent property of the NSE and originates in the latest stages of laminar–turbulent transition controlled by the presence of (rigid) surfaces which is not fully understood yet; for some tantalising clues, though speculative in nature, see Ref. 13 (plus the references to preceding work therein).

Notably, nowadays also the inverse energy cascade (from smaller to larger scales) is a source of present debates in some circumstances (e.g., in the understanding of turbulent particle agglomeration).

Here, we shall not delve deeper in the matter of hydrodynamic instabilities and the transition process in the limit of large values

Table 1. Lower thresholds of Re for turbulence in typical flows.

Flow Case	Re Exceeding	Reference Velocity	Reference Length
Developed circular-pipe and (open-) channel flow	2,400	Flow-area-averaged	Hydraulic diameter
Flat-plate BL	300,000	External-flow speed	Leading-edge distance
Entire BL on circular cylinder in cross flow	10^6	Unperturbed speed	Diameter

of Re; some remarks are given below in connection with boundaries. Our concern is with an already fully developed turbulent flow, i.e., broadband turbulence. This is characterised by the in-time-and-space simultaneous presence of all scales at play (in contrast to the observed dominant/isolated ones governing its history as a transitional flow). Despite the aforementioned inherent difficulties, asymptotic methods provide the proper and powerful means for gaining a deep understanding, at least for the Reynolds- or, equivalently, time-averaged slender shear flow, as they have proven likewise for several decades in the context of laminar and transitional flows. The ultimate goal of this endeavour is to predict the key features and the structure of turbulent flows in the formal limit Re $\to \infty$. Let us note the magnitudes of Re exceeding of which renders developed turbulent shear flows in some important situations and for perfectly smooth surfaces: see Table 1.

1.3. *Governing equations: a critical view*

Although the extremely large shear rates typical of turbulent flows raise doubts whether continuum theory provides the proper framework for their description, this classical approach in connection with the constitutive law for a Newtonian fluid lay the foundations for a most rational turbulence research. This viewpoint is supported by the vast majority of experimental data available, and no convincing arguments have yet led to a strikingly different accepted one. Here, it is noteworthy that relaxation times on a molecular level are still much smaller than the smallest time scales at play and the local Knudsen number $Kn := \hat{l}_p/\hat{l}_t$, formed with the the free path length of the molecules $\hat{l}_p$ and the smallest turbulent (macroscopic) length scale $\hat{l}_p$, is sufficiently small. Then the flow can be safely considered as being locally in thermodynamic equilibrium; the common Stokes hypothesis of vanishing bulk viscosity applies.

Hence, our starting point is formed by the continuity equation

$$\partial_t \rho + \partial_j(\rho u_j) = 0 \tag{1}$$

and the conventional NSE

$$\rho(\partial_t u_i + u_j \partial_j u_i) = -\partial_i p - \rho g_i + \nu_r \left[\partial_{jj} u_i - \frac{\partial_{ij} u_j}{3} \right] \tag{2}$$

for a single-phase Newtonian fluid having uniform dynamic viscosity (which can be regarded a very weak restriction). Here and subsequently, the following conventions and prerequisites are adopted if not stated otherwise. A Cartesian coordinate system and the corresponding covariant Einstein notation are advantageously used $(i, j, k = 1, 2, 3)$; with t being the time, x_i, $u_i(x_j, t)$ and $-g_i$ denote the components of the position vector, the flow velocity, and a body force (especially gravity, most relevant for free-surface flows), respectively, $p(x_j, t)$ is the fluid pressure and $\rho(x_j, t)$ its density; ∂_i (∂_{ij}) indicates first (second) derivatives with respect to x_i (and x_j); all quantities are made appropriately non-dimensional as mentioned above, e.g., u_i, p, ρ with the values taken from the uniform parallel flow assumed infinitely far upstream of a solid obstacle, x_i with a dimension typical of the latter, and t according to x_i and u_i. Furthermore, their dimensional forms are indicated by hats indicate and dependences on Re not stated explicitly.

Thus, any external unsteady external forcing of the flow, imposing naturally one or a multitude of further reference time scales, shall be discarded. However, by the last assumption also the occurrence of free-stream turbulence acting on boundary or free shear layers is excluded, but in many aerodynamic problems of practical relevance the atmosphere around the flight vehicle of interest can in fact be viewed as sufficiently quiet.

The thermal energy equation and an equation of state extend Eqs. (1), (2) to a closed system of governing equations determining the unknowns u_i, p, ρ when supplemented with appropriate BCs, satisfied at walls and for $|x_i| \to \infty$ (upstream), and ICs for $t = 0$, say. However, here our concern is chiefly with Eqs. (1), (2) as we are mostly interested in fundamental properties of turbulent flows which already ensue from the study of incompressible ones of constant density: these are characterised by $\rho \equiv 1$ in Eqs. (1), (2), which thus decouple from and can be treated independently of the energy equation.

We are consequently led to the following self-consistent picture: the flow is assumed to have undergone an intrinsic process of laminar–turbulent transition further upstream due to the presence of solid walls — for external flows, in a boundary layer (BL) emerging for sufficiently large values of Re. Free-stream turbulence causes unstable Klebanoff modes in a shear layer and thus *by-pass* transition, but convectively unstable Tollmien–Schlichting (TS) waves trigger the classical, prevalent routes to shear-layer turbulence on a rather long scale. This scenario due to instabilities (having ubiquitous sources) and/or the receptivity of imposed disturbances, as by surface roughness and/or acoustic waves present in the external flow contrasts with short-scale transition as by reattachment of marginally separated BLs.

Insofar, the presence of walls is crucial, not only for attached turbulent BLs as also granting the existence of free/separated turbulent shear flows as forming nozzles, etc. further upstream. When coinciding with the x_1-axis, say, a wall provokes the usual kinematic no-slip/penetration conditions at

$$x_2 = 0: \quad u_1 = u_3 = 0, \quad u_2 \text{ prescribed}, \tag{3}$$

either by strict impermeability of the rigid wall ($u_2 = 0$) or a (given) rate of suction/blowing. According to Eq. (3) and for slender turbulent shear layers, let subsequently $(x, u) := (x_1, u_1)$, characterise the mean or streamwise flow direction, $(y, v) := (x_2, u_2)$ the one perpendicular to the mean shear or in wall-normal direction, and $(z, w) := (x_3, u_3)$ the spanwise one. Since curvature effects do not have an important influence on such flows, these are neglected from the outset. The coordinates then can also be interpreted as natural ones, e.g., x as tangential to a curved mean-flow streamline or a curved surface, typically a 2D one with generatrices parallel to the z-direction. Such scenarios are tacitly assumed hereafter. They include axisymmetric BL flows; the extension to axisymmetric free jets with x_1 as the axial, x_2 the radial and x_3 the circumferential coordinate is straightforward.

A few words deserve also to be left on the validity of the no-slip condition. It can be argued for liquids by adhesion forces but is less obvious for gases. However, our scales are still so large such that continuum mechanics applies even in the immediate vicinity of the wall. As a consequence of diffuse reflection at a surface, the velocity distribution of the reflected fluid particles becomes statistically independent of that of the incident ones when the spatial scale typical of averaging is much larger than $\hat{l}_p$ but much smaller than $\hat{l}_t$. Therefore, on the latter scale the no-slip condition is observed. For an ideal gas in a state of equipartition, the dynamic viscosity can be expressed as $\hat{\mu} = \hat{\rho}\hat{c}\hat{l}_p/3$ with $\hat{c}$ being the macroscopically observed particle speed, under the basic assumption $Kn = \hat{l}_f/\hat{l}_m \ll 1$ hence the speed of sound, here evaluated at the wall. Furthermore, it is noted that c and ρ undergo changes independent of Re in the x_2-direction. With λ_w denoting the ratio of $\hat{l}_m$ and the global reference length and Ma the Mach number at the reference state, the last condition is then cast in the convenient form

$$\frac{Ma}{(\mathrm{Re}\lambda_w)} \ll 1 \quad \text{for Re} \gg 1. \tag{4}$$

Strictly speaking, it is obligatory to check this criterion in the analysis of BLs driven by external flows at large Mach numbers. However, it is expected to be met throughout for the usually enormous values of Re and λ_w known to be of $O(1/\ln \mathrm{Re})$.[13,14]

1.4. *What exactly is shear-flow turbulence?*

The above introductory view on turbulence motivates the following — for our purposes, adequately complete and precise —

B. Scheichl

characterisation of a turbulent flow. Developed turbulent flows:

(i) Appear naturally for sufficiently large values of Re,
(ii) Are intrinsically stochastic in time and space (keyword: deterministic chaos in mechanical systems) as a consequence of the viscous forces at play howsoever large Re is,
(iii) Are three-dimensional (3D) (associated with vortex stretching).

Items (ii) and (iii) have their origins in the instability scenarios an originally even nominally 2D and steady laminar flow has undergone. Specifically, property (ii) implies that turbulent flows are vortex flows where the dimensions of vortices span the aforementioned range of scales. Issue (iii) is associated with vortex stretching as a typical feature of turbulent flows (we exclude exceptional cases of degenerate 2D turbulence).

In this respect, a typical example of a non-turbulent flow is the famous von Kármán vortex street forming already for $\mathrm{Re} = O(10^3)$ in the wake of plane flow past a closed bluff body (Fig. 1(a)): here, the underlying vortex shedding in the separated shear layers is characterised by a well-defined Strouhal number, and three-dimensionality is still poor, so the typical features of turbulence are not present for such rather moderate values of Re.

 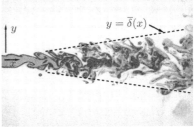

(a) Von Kàrmaǹ's vortex street visualised[12] (© Dr. Sadatoshi Taneda in lieu of The Parabolic Press, modified/reprinted by permission).

(b) Axisymmetric immersed water jet with fractal edge, visualised by laser-induced fluorescence technique[11] (© American Institute of Physics (AIP), modified/reprinted by permission).

Fig. 1. Non-turbulent though highly unsteady (a) versus fully turbulent shear layers (b).

In striking contrast with laminar shear layers having a typical width of $O(\text{Re}^{-1/2})$ as a result of the straightforward analysis of Eqs. (1) and (2), such a definite scaling cannot be given at this stage for developed turbulent ones. As an instructive typical example, the jet depicted in Fig. 1(b) has a numerically small but finite opening angle however large Re becomes when formed with a distance from its virtual origin $x = y = 0$ where it is viewed as fully turbulent $(x = 1)$. This is due to their inertia-driven character and the minor importance of the viscous term in Eq. (2), unlike in the shear-layer balance typical of laminar flows. However, turbulence is in fact in most circumstances confined to a rather slender region having a rather abrupt edge, in the current setting measured by the width $\bar{\delta}(x)$, say. As recognised in Fig. 1(b), the interface separating the turbulent from the entrained (here solely induced) irrotational external flow is a sharp one as the transition towards it not driven by viscous diffusion like in laminar shear layers.

This motivates us to give the:

Definition 1 (slender turbulent shear flow). We speak of a *slender turbulent shear flow* if its width $\bar{\delta}$ satisfies

$$\bar{\delta} \ll 1 \quad \text{and} \quad \frac{\mathrm{d}\bar{\delta}}{\mathrm{d}x} \ll 1, \tag{5}$$

on the global scale x sufficiently far downstream of the transition process.

Albeit quite simple and intuitive, this characterisation proves powerful due to its generality. This allows for remarkable progress by formal asymptotic methods as outlined in Sec. 4. Marked shear $\partial_y u$ across the layer emerges as a pure consequence. Shear flows comprise wall-bounded and separating BLs and free shear flows as separated BLs, jets, mixing layers, wakes behind obstacles and (forced and) buoyancy-driven plumes.

2. On Averaging and Modelling

The classical averaging technique means decomposing the flow into a nominally steady background flow and its turbulent fluctuations

B. Scheichl

about that mean flow, the former studied on the basis of the accord-ingly Reynolds-averaged NSE (RANS). In most cases, the nominal flow is not only steady but also 2D, which represents a decisive sim-plification. It is then independent of z (as mentioned wherever appro-priate in the following).

2.1. *Principles of averaging*

In order to "prove" consistency of the classical closure ideas with the flow structure in the high-Rey-limit in an hitherto unappreciated manner, the basic ingredients to the common averaging strategies have to be condensed.

2.1.1. *Conventional Reynolds-averaging and ergodicity*

By adopting the usual notation, we then decompose any (tensorial) flow quantity, here represented by $\mathcal{Q}$, in the form

$$\mathcal{Q}(x_i, t) = \overline{\mathcal{Q}}(x_i) + \mathcal{Q}'(x_i, t) \quad \text{(2D flow: } \partial_z \overline{\mathcal{Q}} = 0). \tag{6}$$

Here, the mean contribution $\overline{\mathcal{Q}}$ is either interpreted statistically, i.e., in terms of ensemble averaging, or expressed via typical time-averaging:

$$\overline{\mathcal{Q}} := \int_{\mathcal{Q}} \mathcal{Q} \, \mathrm{PDF}(\mathcal{Q}) \, \mathrm{d}\mathcal{Q} = \lim_{\Delta t \to \infty} \frac{1}{T} \int_0^T \mathcal{Q}(x_i, t + \theta) \, \mathrm{d}\theta. \tag{7}$$

Here, PDF stands for the probability density function; more precisely, the first integral in Eq. (7) has to be taken as an Lebesgue measure, and for a stationary process, in the second case implied by the limit process (provided it exists), the celebrated ergodicity theorem (law of large numbers) guarantees the equivalence of both representations. Finally, the relations in Eq. (7) *define* $\mathcal{Q}'$ and give the basic results $\overline{\overline{\mathcal{Q}}} = \overline{\mathcal{Q}}$, $\overline{\mathcal{Q}'} = 0$. We now identify $\overline{\delta}$ in Eq. (5) with the (half) width of the time-mean shear layer. Note that *filtering* the NSE means a (Re-dependent) *finite* filter width T.

As a fundamental finding, $\overline{\mathcal{Q}}$ is interpreted equivalently as expec-tation or time-mean value. Here, Jensen's inequality $\phi(\overline{\mathcal{Q}}) \leq \overline{\phi(\mathcal{Q})}$ for some convex function ϕ is noteworthy, generalising the basic finding $\overline{Q}^2 \leq \overline{Q^2}$. We furthermore recall the nth (statistical) *central moment* $\overline{Q'^n}$ for some integer $n \geq 1$ and note the generalised Cauchy–Schwarz inequality

$$\overline{\prod_{i=1}^{i=n} Q_i} \leq \prod_{i=1}^{i=n} \overline{Q_i^n}^{\frac{1}{n}} \quad (n = 2^m, \quad m \geq 1) \tag{8}$$

involving $2m$ quantities Q_i.

2.1.2. *Favre-averaging*

Classical Reynolds-averaging is generalised by writing

$$Q = \tilde{Q} + Q'', \quad \tilde{Q} := \frac{\overline{\rho Q}}{\overline{\rho}}. \tag{9}$$

Now, Reynolds-averaging equation (1) yields its laminar-like form for steady flow

$$\partial_j(\overline{\rho}\tilde{u}_j) = 0. \tag{10}$$

Hence, this well-known Favre- or density-weighted averaging has proven useful for compressible flows, and we will adopt it subsequently. We accordingly find that $\tilde{\tilde{Q}} = \tilde{Q}$, $\widetilde{Q''} = 0$. By noting that $\overline{\overline{Q}} = \overline{Q}$ and $\overline{\tilde{Q}} = \tilde{Q}$, one obtains $\tilde{Q} = \overline{Q} + \widetilde{Q'}$, $\overline{Q} = \tilde{Q} + \overline{Q''}$, giving the relationship $\widetilde{Q'} = -\overline{Q''} \neq 0$ between the fluctuations Q' and those introduced via Eq. (9).

Favre-averaging contains Reynolds-averaging as special case for vanishing density fluctuations ρ'. This is noticed in connection with some further important relationships, notations and rules explained next.

Let us remind the Steiner translation theorem $\widetilde{Q_1 Q_2} = \tilde{Q}_1 \tilde{Q}_2 + \widetilde{Q_1'' Q_2''}$, that $\widetilde{Q_i Q_j}$ is a *covariance* or *cross-correlation* for $i \neq j$ and a *variance* or *auto-correlation* for $i = j$, and the usual standard deviation $\sigma_Q := \left(\widetilde{Q^2} - (\tilde{Q})^2\right)^{1/2} = \widetilde{Q''^2}^{1/2}$ or RMS value of Q'', being a proper measure for its average magnitude, i.e., the one typical for the predominant fractions of time (where averaged flow quantities can be observed). Hence, by the quite valuable consequence

$$\widetilde{Q_1'' Q_2''} \leq \sigma_{Q_1} \sigma_{Q_2}, \tag{11}$$

of Eq. (8), the once estimated order of magnitudes of the auto-correlations bound those of the cross-correlations.

Moreover, as most important for averaging the governing equations, one readily deduces the following rules: For the time derivatives, $\partial_t \mathcal{Q} = \partial_t \mathcal{Q}' = (\partial_t \mathcal{Q})' = \partial_t \mathcal{Q}''$ such that $\overline{\partial_t \mathcal{Q}} = \partial_t \overline{\mathcal{Q}} = \partial_t \overline{\mathcal{Q}'} = 0$ but $\partial_t Q'' \neq (\partial_t Q)''$, $\widetilde{\partial_t \mathcal{Q}} = \widetilde{\partial_t \mathcal{Q}''} = \widetilde{\partial_t \mathcal{Q}'} = -\widetilde{\partial_t \mathcal{Q}''} = \overline{\rho' \partial_t \mathcal{Q}'}/\overline{\rho} \neq 0$, and

$$\overline{\rho' \mathcal{Q}''} = \overline{(\rho - \overline{\rho})\mathcal{Q}''} = -\overline{\rho}\,\overline{\mathcal{Q}''}\ (\neq 0); \tag{12}$$

for the spatial ones $\partial_i \overline{\mathcal{Q}} = \overline{\partial_i \mathcal{Q}}$, $\partial_i \widetilde{\mathcal{Q}} = \widetilde{\partial_i \mathcal{Q}}$. The latter express the filtering of the small spatial scales/wavelengths characteristic of the turbulent fluctuations, potentially reducing the order of magnitude of the gradients.

Let us next give the following:

Definition 2 (fully turbulent and near-wall regions). Let l_2 and λ_2 denote the locally largest (time-mean or global) scale and the smallest scale (wavelength), respectively, of the turbulent dynamics for the x_2-direction:

The *fully turbulent* regime emerges at scale separation, implying $\lambda_2 \ll l_2$;
The *near-wall* or *viscous sublayer* adjacent to a wall at the collapse $\lambda_2 \sim l_2$.

This just expresses the apparent necessity to distinguish between at least two flow regions in the high-Re limit.

The first regime sufficiently remote from a wall is characterised by

$$|\partial_i \overline{\mathcal{Q}_1' \mathcal{Q}_2'}| \ll |\overline{\mathcal{Q}_1 \partial_i \mathcal{Q}_2'}|, \quad |\partial_i \widetilde{\mathcal{Q}_1' \mathcal{Q}_2'}| \ll |\widetilde{\mathcal{Q}_1 \partial_i \mathcal{Q}_2'}| \quad \text{for Re} \gg 1. \tag{13}$$

Specified for slender shear flows, this yields typical estimates for those being

weakly 2D: $|u'| \ll |\widetilde{u}| = O(1)$, $|v| \ll 1$, $|w| \ll 1$, so $|\widetilde{u_i'' u_j''}| \ll 1$;

$$\tag{14}$$

strictly 2D: $|w'| \ll 1$, $\widetilde{w} \equiv \widetilde{u'' w''} \equiv \widetilde{v'' w''} \equiv 0$. $\tag{15}$

Equations (5), (14) and (15) enable further progress by virtue of a formal asymptotic analysis, by heavy exploitation of Eqs. (11) and (13). The collapse of y-scales in the near-wall regime entails that of the velocity scales, and it turns out that there the classical near-wall scaling applies as a consequence of Reynolds-averaging Eq. (2).

2.2. *RANS and higher-moments transport equations*

Reynolds-averaging of Eq. (2) written in conservative form with the aid of Eqs. (1) and subject to (9) yields the RANS or Reynolds equations

$$\bar{\rho}\tilde{u}_j\partial_j\tilde{u}_i = -\partial_i\bar{p} - \bar{\rho}g_i + \partial_j\bar{\tau}_{ji} + \nu_r\left[\partial_{jj}\bar{u}_i - \partial_{ij}\frac{\bar{u}_j}{3}\right], \quad \bar{\tau}_{ij} := -\overline{\bar{\rho}u_i''u_j''}.$$
(16)

The last double-correlation, equal to $-\overline{\rho u_i'u_j'}$ according to Eqs. (7) and (9), represents the Reynolds stress tensor. Together with Eq. (10) and the correspondingly averaged further governing equations, these equations govern the mean flow but represent an unclosed system, first with regard to the new unknown $\bar{\tau}_{ij}$ — which can be viewed as the central element of what is referred to as the *turbulence closure problem.* Let us recall this intrinsic property of the RANS, where specifying them for incompressible flows ($\rho \equiv 1$, $\rho' \equiv 0$) has Eqs. (10) and (16) decouple from the correspondingly averaged thermal energy equation, so that problem reduces to modelling $\bar{\tau}_{ii}$.

The equations governing the fluctuations ρ' and u_i'',

$$\partial_t\rho' + \partial_j(\bar{\rho}u_j'' + \rho'\tilde{u}_j + \rho'u_j'') = 0,$$
(17)

$$\rho(\partial_t u_i'' + u_j\partial_j u_i'' + u_j''\partial_j\tilde{u}_i) + \rho'\tilde{u}_j\partial_j\tilde{u}_i = -\partial_i p' - \rho'g_i - \partial_j\bar{\tau}_{ji}$$
$$+ \nu_r\left[\partial_{jj}u_i'' - \frac{\partial_{ij}u_j''}{3}\right],$$
(18)

complement Eqs. (10) and (16) to give Eqs. (1) and (2), respectively. We introduce the operator $\mathcal{D}_f$ so as to rewrite the left-hand side of Eq. (18), the contribution of fluctuations to ρ times the total derivative of u_i, as $\mathcal{D}_f\{u_i\}$. Then the expression $\overline{u_j''\mathcal{D}_f\{u_i\} + u_i''\mathcal{D}_f\{u_j\}}$ yields the so-called Reynolds stress equation (RSE) or transport equations for the unclosed terms $\overline{u_i''u_j''}$. More specifically, these are obtained by rewriting Eq. (18) in conservative form with the aid of (1) and taking into account Eq. (12) and Eq. (17) with $\partial_j \overline{(\rho u_j'')} = 0$. We finally have

$$\bar{\rho}(\partial_t + \tilde{u}_k \partial_k) \widetilde{u_i'' u_j''} = R_{ij} + R_{ji}, \quad R_{ij} := P_{ij} + S_{ij}^p + D_{ij}^t + D_{ij}^\nu - \overline{\varepsilon_{ij}^p}. \quad (19)$$

Herein, the tensors at the right-hand side are conveniently distinguished as

turbulent production $P_{ij} := \bar{\tau}_{ik} \partial_k \tilde{u}_j + \overline{u_i''}(\bar{\rho} \tilde{u}_k \partial_k \tilde{u}_j - \partial_k \bar{\tau}_{kj} + \bar{\rho} g_j)$,

$$(20)$$

pressure-shear terms $S_{ij}^p := \overline{p' \partial_i u_j''}$, $\quad (21)$

turbulent diffusion $D_{ij}^t := -\dfrac{\partial_k (\bar{\rho} \widetilde{u_i'' u_j'' u_k''})}{2} - \partial_i \overline{u_j'' p'}$, $\quad (22)$

viscous diffusion $D_{ij}^\nu := \nu_r \left(\dfrac{\partial_{kk} \overline{u_i'' u_j''}}{2} - \dfrac{\partial_i \overline{u_j'' \partial_k u_k''}}{3} \right)$, $\quad (23)$

turbulent dissipation $\varepsilon_{ij}^p := \nu_r \left[(\partial_k u_i'')(\partial_k u_j'') - (\partial_i u_j'')\dfrac{(\partial_k u_k'')}{3} \right]$. $\quad (24)$

It is noted that $\overline{u_i''} \equiv 0$ in Eq. (20) indicates incompressible flow.

Generally spoken, the left-hand side of a transport equations for any averaged quantity exhibits the convective operator $\tilde{u}_j \partial_j$, the right-hand side "diffusive" terms, written in divergence/gradient form, where the viscous ones are those proportional to ν_r, hence further "dissipative" terms also proportional to ν_r, and the remaining "source" or so-called production terms. However, a physical interpretation is only admissible for Eq. (19) as this represents the budget of the share of specific mechanical power exerted by a fluid particle due to the velocity fluctuations u_i'', u_j''. Accordingly, ε_{ij}^p is frequently termed *pseudo*-dissipation: this notation more appropriately matches its physical origin as D_{ij}^ν includes the complementary contribution to dissipation by internal viscous forces (positive by the second law of thermodynamics). Also, typical of these equations are the products of ρ with Favre-averaged terms, as a result of Eq. (9), and that time derivatives vanish identically only in the incompressible-flow limit. However, this is otherwise obtained via Eq. (9) for the sake of consistency with Eqs. (10) and (16) for nominally steady flow.

Equation (19) not only contain triple correlations, cf. Eq. (22), but also double-correlations involving p', cf. Eqs. (21) and (22), and finally such involving only gradients $\partial_i u'_j$, cf. Eq. (24). This reflects the impossibility to formulate a closed systems of equations by considering Eqs. (10) and (16), plus any finite number of arbitrary moments of the NSE, Eq. (2), obtained by multiplying Eq. (18) with terms involving u''_j and subsequent Reynolds-averaging. As an obvious weakness of all types of averaging or filtering Eq. (2), this just means the irretrievable loss of information about the stochastic small-scale dynamics, solely returned in modelled form. Further insight is accomplished, however, by considering the trace of Eq. (19). The resultant budget of the specific turbulent kinetic energy K, the so-called $\widetilde{K}$-equation, serves as the starting point for all considerations on "solving rationally" the closure problem in the limit $\mathrm{Re} \to \infty$:

$$\bar{\rho}(\partial_t + \tilde{u}_j \partial_j)\,\widetilde{K} = R_{ii} = P + S^p + D^t + D^\nu - \overline{\varepsilon^p}, \quad K := \frac{\overline{u''_i u''_i}}{2}. \quad (25)$$

Here, the scalar counterparts to the quantities introduced in Eqs. (20)–(24)

$$P := \overline{\tau}_{ij}\partial_j \overline{u}_i + \overline{u''_i}(\overline{\rho} \tilde{u}_j \partial_j \tilde{u}_i - \partial_j \overline{\tau}_{ji} + \overline{\rho} g_i), \quad (26)$$

$$S^p := \overline{p' \partial_i u''_i}, \quad (27)$$

$$D^t := -\partial_i(\overline{\rho}\,\widetilde{K u''_i} + \overline{u''_i p'}), \quad (28)$$

$$D^\nu := \nu_r \partial_i \left(\partial_i \overline{K} - \frac{\overline{u''_i \partial_j u''_j}}{3} \right), \quad (29)$$

$$\varepsilon^p := \nu_r \left[(\partial_k u''_i)(\partial_k u''_i) - \frac{(\partial_i u''_i)^2}{3} \right], \quad (30)$$

reduce to their well-known standard form in the limit of incompressibility, where Eq. (17) simplifies to $\partial_i u'_i \equiv 0$ ($u'_i \equiv u''_i$).

The structure of Eq. (25) suggests to gain further information by considering $2\overline{(\partial_i u''_j)\partial_i \mathcal{D}_f\{u_j\}}$. A procedure analogous to that leading to the $\widetilde{K}$-equation then yields the so-called $\overline{\varepsilon^p}$-equation, here only specified for incompressible flow for the sake of conciseness:

$$\frac{\overline{u_j \partial_j \varepsilon^p}}{\nu_t} = -2\overline{[(\partial_k u_i')(\partial_k u_j') + \overline{(\partial_i u_k')(\partial_j u_k')}]}\partial_j \overline{u}_i - 2\overline{u_k' \partial_j u_i'}\,\partial_{jk}\overline{u}_i$$
$$+ \partial_i[\partial_i \overline{\varepsilon^p} - \overline{u_i' \varepsilon^p} - 2\overline{(\partial_j u_i')(\partial_j p')}] - 2\overline{(\partial_i u_j')(\partial_i u_k')(\partial_j u_k')}$$
$$- 2\nu_t\overline{(\partial_{ij} u_k')(\partial_{ij} u_k')}. \tag{31}$$

The structure of this equation resembles that of Eqs. (19) and (25). However, only the $\widetilde{K}$-equation is susceptible to a physical interpretation. As exemplifying the associated difficulties categorised above in view of the individual terms in Eq. (31), its last term means a "dissipation of dissipation".

Equations (19) and (25) can be simplified further for shear flows by virtue of Eqs. (11) and (13). A first important conclusion is drawn for firmly attached BLs by inspection analysis, which indicates that the flow regimes introduced by Definition 2 are characterised by

$$\mathrm{Re}^{-1}|\partial_i \widetilde{u}_j| \ll |\overline{\tau}_{ij}|, \quad |D_{[ij]}^\nu| \ll |\varepsilon_{[ij]}^p| \quad \text{for } \mathrm{Re} \gg 1 \tag{32}$$

and
$\mathrm{Re}^{-1}|\partial_i \widetilde{u}_j|/|\overline{\tau}_{ij}| = O(1)$, $|D_{[ij]}^\nu|/|\varepsilon_{[ij]}^p| = O(1)$, respectively. Hence, ν_r enters Eqs. (19) and (25) predominantly via the turbulent dissipation in the fully turbulent region, which enables a drastic simplification of the closure problem for this regime. On the other hand, asymptotic analysis shows that the near-wall time-mean flow exhibits universal properties due to the negligibly small effects of inertia there, which extremely alleviates its treatment; for an outline see Ref. 14 (and the references to pioneering work therein), and Ref. 15. Its scaling is obviously recovered by the classical one,

$$\frac{\widetilde{u}}{\widetilde{u}_\tau} = u^+(x, y^+) = O(1), \quad y^+ := \frac{y}{\overline{\delta}_\nu},$$

$$\overline{\delta}_\nu := \frac{1}{\widetilde{u}_\tau \mathrm{Re}}, \quad \widetilde{u}_\tau := \sqrt{\nu_r \frac{\partial \widetilde{u}}{\partial y}\bigg|_{y=0}}. \tag{33}$$

It is based upon the aforementioned equal order of magnitudes of the molecular and Reynolds shear stresses and the resulting dominant balance of their sum with the wall shear stress. Hence, the local skin-friction velocity $\widetilde{u}_\tau$ provides the suitable velocity scale for the near-wall region.

It is therefore sufficient to restrict the considerations on modelling to the fully turbulent flow. Also, closing $\overline{\tau}_{ij}$ represents the core problem, envisaged next. Closing other relevant quantities then is a subordinate task accomplished in a straightforward manner as far as those enter an asymptotically correct leading order flow description. The central step is to critically review and thereby substantiate the classical Boussinesq hypothesis.

2.3. *A promising view on the Boussinesq ansatz*

Let us apply the common decomposition of $\partial_i u_j$ into its symmetric and anti-symmetric part, i.e., the rate-of-deformation or strain-rate tensor $S_{ij} := (\partial_i u_j + \partial_j u_i)/2$ accounting for stretching and volumetric dilatation of the particles, and the rotational contribution $\Omega_{ij} := (\partial_i u_j - \partial_j u_i)/2$ by the vorticity, $\varpi := \epsilon_{ijk}\partial_j u_k = \epsilon_{ijk}\Omega_{ij}$ (Levi-Cività symbol ϵ_{ijk}). Now consider a tensor Σ_{ij} depending on the local state of deformation of the fluid considered, i.e., the velocity gradient $\partial_i u_j$ and possibly higher derivatives in an Eulerian frame of reference. As a fundamental finding of continuum mechanics,[16] two consecutive kinematic statements can be made on a sole isotropic dependence on $\partial_i u_j$, i.e., one invariant against reflections and rotations of the coordinate system (Kronecker delta δ_{ij}):

Theorem 1 (Rivlin–Ericksen representation theorem).

(a) *Isotropy requires Σ_{ij} to be a function of S_{ij} solely;*
(b) *for a symmetric tensor $\Sigma_{ij} \equiv \Sigma_{ji}$ having the irreducible invariants $I_1 = S_{ii}$, $I_2 = (S_{ii}S_{jj} - S_{ij}S_{ij})/2$, $I_3 = \det S_{ij}$, any isotropic dependence on S_{ij} is of the generic quadratic form*

$$\Sigma_{ij} = \nu_0\,\delta_{ij} + \nu_1 S_{ij} + \nu_2 S_{ik}S_{kj} \quad with \quad \nu_{0,1,2} = \nu_{1,2,3}(I_j), \quad (34)$$

being some so-called structure functions.

If Σ_{ij} is identified with the Cauchy stress tensor (symmetry is entailed by the Boltzmann axiom), the linear two-parameter constitutive law for a Newtonian fluid represents the simplest conceivable prototype of such a relationship (here $\nu_1 = (\nu_b - 2\nu_r/3)S_{ii}$ with ν_b being the (kinematic) bulk viscosity, $\nu_2 = 2\nu_r$, $\nu_3 = 0$). Bearing this in mind, the question arises to which extent the deep interrelation (34) allows for a putative dual by setting $\Sigma_{ij} = -\widetilde{u_i'' u_j''}$. The following considerations on the crucial issues of such a relationship (locality, stress–strain-type relation, linearity, local isotropy) guide us in establishing a Reynolds-stress closure in this spirit. This is then found fully consistent with the characteristics of high-Re turbulence.

Locality. This might be questionable at a first glance, given upstream influences on the flow and its history. However, a turbulent shear flow adjusts quite rapidly to local conditions as long as changes in the external-flow conditions or the surface topography are sufficiently smooth. Contrarily, short-scale disturbances (typically, by sudden changes in the wall roughness, individual wall-mounted obstacles topography or shock-impingement on a BL) provoke a distinctly slow recovery of the flow. However, even this phenomenon can be traced back merely to inertial effects, inasmuch as the anticipated scale separation underlying the modelling is eradicated only locally. Moreover, a turbulent flow "forgets" rather quickly the particular mechanism of laminar–turbulent transition as it manifests itself in a generic manner independent of this and, like in laminar flow, it is the ellipticity of the NSE which accounts for further non-local effects non-locality. Finally, the extensions of Theorem 1 including history effects account for viscous relaxation but in Newtonian turbulent flows these are purely inertia-driven. As a conclusion, locality in the Reynolds stress–strain relationship does not pose a seriously troublesome issue.

A pure stress–strain relationship. In fact, any dependence of $\overline{\tau}_{ij}$ on higher order gradients of $\tilde{u}_i$ would increase the order of the RANS compared to that of the underlying NSE and thus raise an inconsistency, mostly in terms of the BCs given by Eq. (3) to be satisfied.

Linearity. As for locality, nonlinear inertia terms are an essential feature of the RANS, so that ν_2 in Eq. (34) is virtually set to zero.

Then the nonlinearities involving the small-scale motion are coped with though by the proper modelling of ν_0 and ν_1, which then must not depend on the (averaged) invariants $\widetilde{I}_1$, $\widetilde{I}_2$, $\widetilde{I}_3$ for the sake of consistency. As the original NSEs have only quadratic nonlinearities and for consistency with the preceding point raised, these structure functions are required to depend on double-correlations involving first derivatives of u_i'' solely.

Isotropy. In the high-Re limit and for the associated smallness of the smallest scales identified in a turbulent flow, Kolmogorov's hypothesis of *local isotropy*[17] seems reasonable. Here, this is put in a different context: the relationship Eq. (34) for $\Sigma_{ij} = \overline{\tau}_{ij}/\overline{\rho}$ is a locally isotropic one but not the tensor $\overline{\tau}_{ij}$ itself (which would imply the much more restrictive global isotropy). As of great practical value, we cast this in the following Proposition.

Proposition 1 (local isotropy). *In the limit* $\mathrm{Re} \to \infty$, *for a turbulent slender shear layer all components of* $\overline{\tau}_{ij}$ *scale in the same manner [equality in (11)] : some gauge function* $\gamma(\mathrm{Re}; y)$ *gives* $\lim_{\mathrm{Re}\to\infty} \overline{\tau}_{ij}/\gamma = O(1)$.

By recalling the hydrostatic stress contribution $\overline{\tau}_{ii}/3 = -2\overline{\rho}\widetilde{K}/3$, we arrive at the well-known Boussinesq ansatz for the deviatoric stresses:

$$\overline{\tau}_{ij} = \nu_0\,\delta_{ij} + \nu_1 \widetilde{S}_{ij}, \quad \nu_0 = -\frac{2(\widetilde{K} + \nu_t \partial_i \widetilde{u}_i)}{3}, \quad \nu_1 = 2\nu_t. \tag{35}$$

This is seen as the "turbulent" counterpart to the phenomenological relationship for a Newtonian fluid. It introduces the so-called (kinematic) eddy or turbulent viscosity ν_t. In accordance with the above considerations, it is assumed to depend isotropically on the averaged motion on the microscopic scales (the turbulent fine structure) and on its molecular counterpart (herewith on the thermodynamic state), the perturbation parameter ν_r, solely. The kinematic dependence then can only involve $\widetilde{K}$ and $\overline{(\partial_i u_j'')(\partial_i u_j'')}$ in the limit $\mathrm{Re} \to \infty$. Dimensional analysis gives

$$\frac{\hat{\nu}_t\,\overline{\widetilde{\varepsilon^p}}}{\widehat{\widetilde{K}}^2} \sim \Pi\left(\frac{\hat{\nu}_t}{\hat{\nu}_r}\right). \tag{36}$$

From here on, density variations are neglected. Furthermore, the estimates in Eq. (32) characteristic of the fully turbulent region imply $\nu_t/\nu_r \to 0$ as $\text{Re} \to \infty$ there, so that the function Π in Eq. (36) tends to a constant c_ν, say. This rationale recovers the widely-accepted, deceptively simple formula for the fully turbulent flow regime, as an asymptotic one on a sound basis:

$$\nu_t \sim \frac{c_\nu \overline{K}^2}{\overline{\varepsilon^p}}(c_\nu \approx 0.09) \quad \text{for Re} \gg 1. \tag{37}$$

Hence, the empirical value of c_ν is an asymptotic property of the NSE.

In analogy to the above dimensional considerations, one may alternatively introduce a turbulent length scale measuring the diameter of the largest eddies, the so-called mixing length ℓ to be attributed to Prandtl.[18] With his choice of a new constant c_P motivated by empirical observations, one then writes

$$\nu_t \sim c_P \ell \sqrt{\overline{K}} \quad (c_P := 0.55). \tag{38}$$

This is fully equivalent with Kolmogorov's similarity hypothesis[17] whereby $\hat{\nu}_t$ is expressed as product of a length and velocity typical of the turbulent motion, given by $\hat{\ell}$ and $\widehat{K}^{1/2}$, respectively. Finally, combining Eqs. (37) and (38) yields the famous (experimentally confirmed) Prandtl–Kolmogorov formula.[3,18]

$$\ell = \frac{c_\varepsilon \overline{K}^{\frac{3}{2}}}{\overline{\varepsilon^p}} \left(c_\varepsilon := \frac{c_\nu}{c_P} \approx 0.168 \right). \tag{39}$$

Notably, at this stage no assertions on the variations of the quantities ν_t, $\overline{K}$, $\overline{\varepsilon^p}$, ℓ with Re are made.

Most of the commonly adopted turbulence closures rely on the Boussinesq ansatz, or in combination with one of Eqs. (37)–(39) as outlined next.

2.4. Categorisation of closures

Available closures are (roughly) divided into the following different families. For their most salient and widespread members, we refer to Refs. 7–9; for shear layer approximations according to Eqs. (13),

(14), (32) to Ref. 3. Here, Eq. (16) reduces to the least-degenerate form

$$\bar{\rho}(\tilde{u}\,\partial_x\tilde{u} + \tilde{v}\,\partial_y\tilde{u}) \sim -\partial_x\bar{p} - \bar{\rho}g_i + \partial_y\bar{\tau}, \quad \partial_y\bar{p} \sim 0, \quad \bar{\tau} := \bar{\tau}_{12} \sim \nu_t\,\partial_y\bar{u}. \tag{40}$$

2.4.1. *Incomplete closures*

Incomplete models rely on directly formulating algebraic expressions or ordinary differential equations (ODEs) with respect to x either for ν_t or ℓ.

Algebraic (zero-equation) closures. These simplest models only resort to the continuity and Reynolds equations, Eqs. (10) and (16) as here ν_t is ad hoc expressed in terms of $\overline{S_{ij}}$. Insofar, these are essentially nonlinear and *non-rational* models. This is so because such an approach does not necessarily comply with the constitutive-type relationship (37) but also as they predict a physically unacceptable, diffusive transition from the turbulent to the external, mainly irrotational flow. The most popular eddy-viscosity-based algebraic closures are the Cebeci–Smith and Baldwin–Lomax models.

The classical picture of turbulent BLs predicts a predominantly irrotational flow in their fully turbulent region with the weak vortical perturbations associated with the Reynolds shear stress (cf. Sec. 4.2.1). Irrotational mean-flow components in the i-th direction are described by $\partial_i\varphi$ with a steady scalar potential φ. As Eq. (10) then gives $\partial_{ii}\varphi = 0$ for incompressible flow and the mean shear rates read $\overline{S_{ij}} = \partial_{ij}\varphi$, we have $\partial_j\bar{\tau}_{ji} \propto \partial_{ijj}\varphi = 0$ in Eq. (16), in agreement with that flow structure. Indeed, its consistency with Eqs. (35) and (36) is also guaranteed by the higher order models below when exposed to the underlying asymptotic expansions.

One-equation closures. Equation (37) suggests to supplement the basic equations Eqs. (10) and (16) with the transport equations (19), (25) and (31) for $\bar{\tau}_{ij}$ and, as entering Eq. (37), the turbulent kinetic energy and dissipation, respectively. Transport equations for triple-correlations are neglected (in view of the quadratic nonlinearities of the NSE). In the aforementioned equations, they together with correlations involving p' contribute to the turbulent diffusion terms and are modelled as such ad hoc in analogy to their viscous counterparts: in the closed transport equation for some passive scaler $\overline{Q}$ (convected with $\overline{u}_i$), they become $\partial_i[(\nu_t/Pr_Q)\,\partial_i\overline{Q}]$ with some turbulent Prandtl number Pr_Q forming a further empirical input, often assumed as constant. For instance, closing D^t in Eqs. (25) and (28) in this manner by identifying Q with $\overline{K}$ gives

$$\overline{u}_j\partial_j\overline{K} = \overline{\tau}_{ij}\partial_j\overline{u}_i + \partial_i\left[\left(\frac{\nu_r + \nu_t}{Pr_{\overline{K}}}\right)\partial_i\overline{K}\right] - \frac{c_\varepsilon\overline{K}^{\frac{3}{2}}}{\ell} \tag{41}$$

for incompressible flow. One-equation closures only employ Eqs. (10) and (16), and the modelled $\overline{K}$-equation (41) as ℓ is modelled independently.

The algebraic mixing-length closures, as widely employed for BL calculations, are, correctly speaking as demonstrated in Sec. 4.1, asymptotically reduced one-equation closures. These in addition adopt the $\widetilde{K}$-equation. The same holds for the refinements involving ODEs, which are accordingly often referred to as "one-half"-equation or "one-one-half"-equation closures. A popular, interesting member of this class is the Johnson–King "non-equilibrium" model, which seeks $\max\overline{\tau}(x)$ across the BL. To this end, in its fully turbulent main portion one expresses $\widetilde{K}$ as proportional to $\overline{\tau}$ in the $\widetilde{K}$-equation to obtain $\nu_t = \max(\overline{\tau}/\partial_y\overline{u})(x)$; cf. Eqs. (40) and (41).

2.4.2. Complete closures

Complete models also adopt a modelled form of the $\overline{\varepsilon^p}$-equation (31) or, equivalently, of the transport equation for some scalar

$$Z := \overline{K}^q\ell^r \propto \overline{K}^{q+\frac{3r}{2}}\,\overline{\varepsilon^p}^{-r}, \quad q, r \in \mathbb{Q} \quad \text{(otherwise largely arbitrary)} \tag{42}$$

or even the RSE, Eq. (19), as outlined above: no longer specific models for ν_t or ℓ are required.

Two-equation closures.   One formally obtains the model equation for Z by means of the following "recipe":

(1) "multiply" Eq. (41) with $Z/\overline{K}$ while differential operators are ignored;

(2) replace the arising factor $\overline{\varepsilon^p}/\overline{K}$ by $\overline{K}/\nu_t$, according to Eq. (37), so as to model consistently the dissipative term as e.g., the last one in Eq. (31);

(3) insert proportionality constants and a further turbulent Prandtl number at the appropriate places.

This yields

$$\overline{u}_k \partial_k Z = C_{P,Z} \left(\frac{Z}{\overline{K}} \right) \overline{\tau}_{ij} \partial_j \left[\overline{u}_i + \partial_i \left(\frac{\nu_r + \nu_t}{Pr_Z} \right) \partial_i Z \right] - \frac{C_{\varepsilon,Z} \overline{K}}{\nu_t},$$

(43)

containing the two model constants $C_{P,Z}$, $C_{Z,\varepsilon}$ representative for the arising production and dissipation terms and the Prandtl number Pr_Z to be modelled as accounting for the associated turbulent diffusivity. The most pervasive combinations of n and m, namely leading the commonly adopted two-equation closures, are given in Table 2. (In Rotta's ℓ-model, ℓ is redefined as the "natural" length scale $\ell := \int_0^\infty \overline{u_i'(x_j, t) u_i'(x_j + \xi_j, t)} \, d\xi_j / (2\overline{K})$.)

Three-equation or Reynolds-stress equation closures. Finally, the so-called three-equation closures even discard the Boussinesq ansatz in favour of supplementing Eqs. (25) and (31) with the six scalar RSE, Eqs. (19) and (24), now truncated as $u_i' \equiv u_i''$, $\partial_i u_i' \equiv 0$:

$$-\overline{u}_k \partial_k \overline{\tau}_{ij} = \overline{\tau}_{ik} \partial_k \overline{u}_j + \overline{\tau}_{jk} \partial_k \overline{u}_i + S_{ij}^p + S_{ji}^p + D_{ij}^t + D_{ji}^t - \nu_r \partial_{kk} \overline{\tau}_{ij} - 2\overline{\varepsilon_{ij}^p}.$$

(44)

Table 2. Most popular two-equation models (see Eq. (42) for q, r).

q	r	Denotation of Z	$\varepsilon^p \propto$	Authors (see Refs. 3, 7–9)
3/2	−1	ε^p	Z	Chou, Jones, Launder
1/2	−1	ω	$Z\overline{K}$	Kolmogorov, Wilcox
1	−2	ω^2	$z^{1/2}\overline{K}$	Spalding
1	1	n/a	$Z^{-1}\overline{K}^{5/2}$	Rotta, Spalding
0 (1)	1	ℓ $(\overline{K}\ell)$	$Z^{-1}\overline{K}^{3/2}$ $(z^{-1}\overline{K}^{5/2})$	Rotta

Here, the remaining unclosed terms S_{ij}^p, D_{ij}^t, $\overline{\varepsilon_{ij}^p}$ are consequently modelled in terms of $\overline{\tau}_{ij}$, $\overline{K}$, $\overline{\varepsilon^p}$ as follows. At first, we note that the pressure fluctuations give non-symmetric contributions to S_{ij}^p, but only symmetric tensors of 12 model parameters $c_p^{ij} = c_p^{ji}$ and $Pr_\tau^{ij} = Pr_\tau^{ji}$ are introduced in

$$S_{ij}^p + S_{ji}^p = -c_p^{(ij)}\,\overline{K}\,\overline{S_{ij}}, \quad D_{ij}^t + D_{ji}^t = -\partial_k\left[\left(\frac{\nu_t}{Pr_\tau^{(ij)}}\right)\partial_k\overline{\tau}_{ij}\right]. \quad (45)$$

A further tensor of six constants $c_\varepsilon^{ij} = c_\varepsilon^{ji}$ is needed to express

$$\overline{\varepsilon_{ij}^p} = \frac{\delta_{ij}\overline{\varepsilon^p}(1-2c_\varepsilon^{(ij)})}{3} - \frac{c_\varepsilon^{(ij)}\overline{\varepsilon^p}\,\overline{\tau}_{ij}}{\overline{K}}, \quad (46)$$

such that taking the trace of Eq. (44) now recovers twice Eq. (41).

Any potential benefit of avoiding the Boussinesq hypothesis in the RSE closure is definitely impaired by the myriad of adjustable model constants.

2.5. *Some critical aspects*

Some remarks shall be devoted to two popular extensions of the two-equation models. At first, models involving nonlinear extensions of the linear Boussinesq hypothesis in terms of an explicit dependence of ν_t on $\widetilde{S}_{ii}$ and $\widetilde{I}_2$ have gained awareness. The Menter two-equation or shear stress transport (SST) model pertains to this family as the currently probably most relevant representative. Here, the BL approximation of a shear-dependence of ν_t shall improve modelling of the blending between the fully turbulent and the near-wall region. Secondly, there are specific features concerning modelling the terms that account for compressible effects. The Menter baseline (BSL) model is a quite popular member of the class of models coping with strongly compressible flows. Also, including compressibility in the eddy-viscosity closure has attracted attention. However, this has led to numerical instabilities and no definite enhancement over the standard BSL model been achieved so far (here the last word has not been spoken yet).[8] Both the SST and the BSL models have proven superior over the classical two-equation closures and highly successful for flows in a wide range of engineering applications, in particular, such undergoing gross separation. They are reviewed in Ref. 9 and still developed further.

Non-locality in also accounted for by considering correlations $\widetilde{\mathcal{Q}_1'' \mathcal{Q}_2''}$ where $\mathcal{Q}_1''$, $\mathcal{Q}_2''$ are calculated at different positions x_i. However, establishing reasonable closures of the so arising two-point correlations has proven a much less viable task compared to modelling of the conventional one-point-correlations (which originate in the locality of the NSE).

Despite these more recent modelling activities, the popularity and undeniable success of the classical Boussinesq formulation, Eq. (35), for high-Re turbulence resorts to its theoretical foundation, revealed in Kolmogorov's, Prandtl's and coworkers' seminal work as forerunners. The often mentioned shortcoming of this ansatz when applied to even conventional BL flows but with sudden changes of external conditions, as addressed in the context of locality above, is critical at least locally where Eq. (35) predicts zero Reynolds shear stress at zero mean shear rate, i.e., for local maxima of $\bar{u}$. This is definitely in doubt as in conflict with experimental evidence. The same situation emerges at the onset of backflow by mean separation. However, since the associated failure of the *rationally* founded linear model is tied in with the local breakdown of the cascade of disjunct scales anticipated by Eqs. (13) and (14), attempts to establish *non-rational* nonlinear ones cope only insufficiently with it. We elucidate this next.

3. Comments on "Turbulence Asymptotics"

The central challenge in a fully rational (model-free or *ab initio*) treatment of high-Re turbulence lies in the simultaneous presence of all scales.

Let us consider Eqs. (17) and (18), which describe the fluctuations provided the mean field is given. Usual multiple-scales and homogenisation techniques are appropriate for the asymptotic treatment of several spatial/temporal scales as long as a hierarchy of problems can be established and the dependence of some quantity on the small scales entails a solvability condition regarding the long-scale behaviour in a lower order approximation. However, such an approach must fail in broadband turbulence as the coefficients in the associated equations resort in the background flow: dependences on global scales involve in all approximations the averaged dependence on all the smaller ones, which requires a sophisticated method of *beyond-all-order* asymptotics — which is not available.

The best one can do here is to resort to Eq. (13) at each level of approximation, expressing the equivalence of complete time-averaging and filtering all scales but the global ones. However, this is not so little. One just must be aware of a specific peculiarity: order of magnitudes might be reduced by averaging, so that averaged equations, containing all scales, are "fuller" than the individual ones that arise by expanding the NSE. An associated theoretical framework,[19] however, has not proven convincingly superior compared to the asymptotic concept pursued here and based on Definition 1, Proposition 1, Eqs. (13)–(15), (32) and (33) and the scaling arguments concerning the fully turbulent part of BLs given in Sec. 4.2.1.

Now let us refine the ideas of exploiting the unsteady-flow scaling: Because of Eq. (32) and since $\sigma_{u_i'} = O(\gamma^{1/2})$ by Proposition 1 and $\overline{\varepsilon^p} = O(1)$ at the maximum in Eq. (25), the possibly smallest scales at play for most of the time are of $O(\sqrt{\nu_r \gamma})$. Since the presence of a wall and thus $\widetilde{u}_\tau$ in Eq. (33) provides a lower bound of the amplitudes $\sigma_{u_i'}$ associated with the smallest scales in the fully turbulent regime, there the viscous term in Eq. (18) of $O(\gamma^{-1/2})$ is comparatively small. These considerations hold also for $\varepsilon^p = o(1)$. Hence, the leading order approximation of Eq. (18) just implies a Rayleigh stage governing those smallest scales carried by the mean flow,

$$\partial_t u_i' + \overline{u}_j \partial_j u_i' \sim -\partial_i p' \quad \left[= O\left(\sqrt{\frac{\gamma}{\nu_r}} \right) \right], \tag{47}$$

(compressibility has no significant effect here). Thus, self-sustained turbulence means that the unstable Rayleigh waves are damped on the associated longer scales. That is, for $\overline{u}_i$ taken as prescribed, a formal multiple-scales approach might deepen our insight into the dynamics of the fluctuations, at least for the aforementioned portions of t where the averaged equations allow for identifying distinct scales. This ties in well with the (intensely debated) doubts on the validity of unsteady-BL theory[20] as severe small-scale mechanisms are potentially overlooked in a shear layer setting: the suggested "race" between modal instabilities (TS and Rayleigh stage) and inherently nonlinear ones leading to blow-ups (TS scale) loses its criticality for finite values of Re, damping those sufficiently. Filtering such small scales finally provokes an ill-posedness of the equations governing the longer-scale dynamics, but this does not render the overall asymptotic approach invalid. For a recent discussion, see Braun & Scheichl[21] (triple-deck/TS scales), Cassel & Conlisk.[21] In case of developed turbulent flow, finally the impossibility manifests to determine the mean flow on the largest scales $[x = O(1), y = O(\bar{\delta})]$ in a finite number of steps, i.e., a hierarchical asymptotic concept.

The last statement agrees with another feature typical of turbulence:

Observation 1 (non-interchangeable limits). A solution of Eqs. (1) and (2) (and appropriate ICs/BCs) does in general *not* converge to that of the corresponding Euler equations (and the identical ICs/BCs) as Re $\to \infty$.

In fact, rather the truncated Euler equations (47) hold. A prominent candidate for an exception are classically scaled (firmly attached) turbulent BLs in the fully turbulent main portion of which the steady, imposed potential flow predominates (Sec. 4.2.1). As Ω_{ij} is dual to the vorticity, one finds $\overline{\varpi'^2} \equiv 2\,\overline{\Omega'_{ij}\Omega'_{ij}} \equiv \overline{(\partial_i u'_j)(\partial_i u'_j)} - \overline{(\partial_i u'_j)(\partial_j u'_i)}$. The last term equals $-\partial_{ij}\overline{u'_i u'_j}$ with the aid of Eq. (17), thus it is negligibly small by Eq. (13). By these expressions, the Helmholtz's vortex theorem not only prevents the generation of vorticity in an inviscid flow but any small-scale fluctuations. This simple but nonetheless outstanding finding corroborates the splitting into two mainly inviscid flow regions initiating scale separation: an external, predominantly irrotational and fluctuation-free one, reigned by the full Euler equations, and a turbulent shear layer, characterised by Eqs. (5) and (47), and crucially Re-dependent small scales.

4. The Asymptotic Framework of Turbulent Shear Flows

Before skipping down to a topical insight into turbulent BLs, we envisage free shear layers. These are easier to deal with as they lack wall binding.

4.1. Free slender shear flows reappraised

For the original asymptotic analysis of free shear layers (and the associated near-field close to a nozzle, etc.), solely resorting to (14) and the empirical finding (5), we refer to Ref. 22, for a review cf. Ref. 3.

Due to the absence of wall binding, the viscous term in Eq. (16) is of subordinate importance across the whole shear layer. In turn, ultimately developed turbulence means Re-independence of $\overline{\tau}$ as $Re \to \infty$ and Eq. (40) retained in full (apart from the body-force term). We are therefore concerned with an *ad hoc* scaling $\overline{\delta} = O(\alpha)$, $\overline{\tau} = O(\alpha)$ where the small but finite so-called slenderness parameter α represents the principal perturbation parameter aside from Re. Our earlier analysis gives $\overline{K} = O(\alpha)$, $\sigma_{u_i'} = O(\alpha^{1/2})$, hence γ is identified with α in the main portion of the shear layer according to Proposition 1. We then have $P \sim \overline{\tau}\partial_y\overline{u} = O(1)$ in Eq. (25), and the only candidate to enter the so reduced $\overline{K}$-equation apart from $\overline{\varepsilon^p}$ is the dominant approximation of D^t, here $-\partial_y\overline{v'p'}$, see Eq. (28). As this quantity drops out by integration across the shear layer, $\varepsilon^p = O(1)$ is confirmed and the bracketed scaling in Eq. (47) applies. Moreover, Eqs. (1) and (2) give $-(\partial_i u_j)(\partial_j u_i) \equiv \Omega_{ij}\Omega_{ij} - S_{ij}S_{ij} = \partial_{ii}p$, posing a "Poisson problem" for p' to leading order subject to homogeneous boundary conditions imposed at the edges of the turbulent region. This yields $p' = O(\alpha)$, $D^t \sim -\partial_y\overline{v'(K + p')} = O(\alpha^{1/2})$ rather than $p' = O(\alpha^{1/2})$, $D^t = O(1)$, which reduces Eq. (25) finally to the balance

$$\overline{\tau}\partial_y\overline{u} \sim \overline{\varepsilon^p} \; [= O(1)]. \qquad (48)$$

Townsend coined the notion *structural equilibrium*[2] for shear flows where Eq. (48) applies. From the asymptotic viewpoint, it only holds for a large velocity deficit with respect to the non-turbulent region, i.e., for $\partial_y \bar{u} = O(1/\bar{\delta})$, although frequently also assigned to the overlap between the fully turbulent and the near-wall region in BLs.[3] However, diffusion is at play both in the latter region, where Eq. (25) stays fully intact, and in the particular layer on its top in a multi-tiered BL.

Noticing the shear layer approximation $\tau \sim \nu_t \, \partial_y \bar{u}$ of Eq. (35) and substituting Eqs. (38) and (39) into Eq. (48) reveals the mixing-length closure

$$\nu_t \sim (c_\ell \ell)^2 |\partial_y \bar{u}|, \quad \bar{\tau} \sim (c_\ell \ell)^2 \partial_y \bar{u} |\partial_y \bar{u}| \quad \left(c_\ell := \frac{c_P}{c_\nu^{1/4}} \approx 1.004 \right).$$

(49)

Considering the exchange of turbulent momentum across mean shear on a "mixing" length scale ℓ yields this expression for ν_t independently.[18] Hence, algebraic mixing-length closures for $\bar{\tau}$ are seen as asymptotically correct one-equation closures. This underpins their strength and the recommendation to discard typical eddy-viscosity closures in favour of suitably modelling an $O(1)$-function l as $\ell/\bar{\delta}(x) \sim \alpha^{1/2} \, l(\eta)$ with $\eta := y/\bar{\delta}$. One typically finds

$$\alpha \approx 0.1 \quad \text{(free shear layers)}, \quad c_l := \alpha^{\frac{1}{2}} \approx 0.085 \quad \text{(BLs)}. \quad (50)$$

This completes the empirical input by proposing fixed real-world values α should take on. We remark that all commonly used algebraic mixing-length closures employ a value of c_l close to that in Eq. (50), but the most popular ad hoc (diffusive) eddy-viscosity-based algebraic Cebeci–Smith and Baldwin–Lomax models propose $\alpha = 0.0168$.[3] As a most salient result, we find that $\bar{\delta} \gg \ell = O(\alpha^{3/2})$ and $\nu_t = O(\alpha^2)$, but this is only compatible with Eqs. (36)–(38) if the constants therein are taken as of $O(1)$. However, the value of c_ν is reliable for BLs but definitely larger for free shear flows, so the numbers in Eq. (50) can be confidently viewed as small.

Assuming $l(1) > 0$ predicts a physically reliable abrupt edge of the shear layer and implies $\bar{\tau} = \partial_Y \bar{\tau} = \partial_Y \bar{u} = 0$ there and thus continuously differentiable flow quantities there. This allows for a sufficiently smooth patching with the external flow. This obviates the need to consider a passive overlayer (of width $\alpha^{3/2}$ as there convection and diffusion are retained in Eq. (25)) that accounts for intermittency.[22] The latter can be taken care of properly though by multiplying the formula for ν_t in Eq. (49) with well-known Klebanoff's empirical intermittency probability function $I_K(\eta) := 1/(1 + 5.5\,\eta^6)$,[3] which improves the computation of the flow near the shear-layer edge and lessens the otherwise often over-predicted values of $\overline{\tau}$ to more realistic ones.

As $\overline{\tau}$ changes sign in a free shear layer, let this take place at $\eta = 0$ (axis of free jet, see Fig. 1(b), or dividing streamline in a mixing layer). There also P vanishes, $\overline{\tau}$ varies linearly with η, and $\overline{u}$ must be regular. As ℓ in Eq. (49) would behave singularly there, an inner region is necessitated where also D^t enters Eq. (48). From Eqs. (28), (38) one infers that $\overline{\varepsilon^p} = O(\alpha^{3/2}/\ell)$ and $D^t = O(\alpha^{3/2}/y)$, so that the inner layer emerges for $\eta = O(\ell/\alpha)$ or $y = O(\overline{\delta}_i)$, say. There $\overline{\tau} = O(\overline{\delta}_i)$ implies $\partial_y \overline{u} = O(\overline{\delta}_i^{-1/2})$. Because $\partial_y \overline{u}$ must vary with η^a where a is some constant, say, in the overlap conjoining both layers, there l must vary with $\eta^{3/2-a}$ sub-linearly $(a > 1/2)$; see Eq. (49). The $\overline{K}$-equation does not provide any further information here. However, Prandtl's original mixing-length concept outlined above Eq. (38) is of avail here, which then does not change its order of magnitude in a free shear flow as a wall is absent: $r = 0$, i.e., $\ell = O(\alpha^{3/2})$ in the entire layer, giving $\overline{\delta}_i = O(\alpha^{3/2})$ and $\gamma = O(\alpha^{3/4})$ in its inner part: Fig. 1(b), Table 3.

It is physically evident that the freely moving large eddies determine the inner length scale, entering the mean-flow description where Eq. (48) degenerates, e.g., due to symmetry of a jet flow with respect to its centreline. There $\overline{u}$ exhibits a cuspidal singularity (perturbing the finite centreline speed), so that the $\overline{u}$-variations

Table 3. Scaling of $O(1)$-deficit shear flow summarised (Landau symbols omitted).

	Thickness	ℓ	ν_t	$\tau_{ij},\ \overline{K}$	$P,\ \overline{\varepsilon^p}$	D^t	$Pr_{\overline{K}}$
Outer (main) layer:	α	$\alpha^{3/2}$	α^2	α	1	$\alpha^{1/2}$	$\alpha^{3/2}$
Inner (diffusive) layer:	$\alpha^{3/2}$	$\alpha^{3/2}$	$\alpha^{9/4}$	$\alpha^{3/2}$	$\alpha^{3/4}$	$\alpha^{3/4}$	1

are of $O(\bar{\delta}_i^{1/2})$ in the inner layer. There turbulent diffusion accounts for the aforementioned regular behaviour of all flow quantities and Eqs. (36)–(38) hold as well; for modelling aspects, see Ref. 22. It is noted that α is also a measure for the (Re-independent) entrainment of the flow and $Pr_{\overline{K}}$ must be taken as small in the main layer.

Incompressibility of the shear flow is effectively allowed by its slenderness to the asymptotic accuracy considered. By introducing a streamfunction ψ to satisfy Eq. (10), we then write for both free layers and BLs

$$[\bar{u}, \bar{v}] = [\partial_y \psi, -\partial_x \psi], \quad [\psi, \bar{\tau}] = [u_r(x)\,\bar{\delta}(x)f(x, \eta),\, u_r(x)^2 s(x, \eta)]. \tag{51}$$

Herein, the reference speed u_r of $O(1)$ is either that along the centre-line (jet, pressure-free) or that imposed by an external potential flow (wake, mixing layer, BL), in the latter case subsequently denoted as u_e. Expanding

$$[f, s] \sim [F, \alpha S] + \cdots, \quad \bar{\delta} \sim \alpha\, \Delta(x) + \cdots, \tag{52}$$

yields the leading order shear-layer approximation of Eq. (16) for free flows:

$$(\mathrm{d}_x \ln u_r)(F'^2 - I) - [\mathrm{d}_x \ln(u_r \Delta)]FF'' + F'\partial_x F' - F''\partial_x F = \frac{S'}{\Delta}. \tag{53}$$

Here and hereafter in this context, d_x stands for x-derivatives and primes for η-derivatives. The indicator function I equals 1 if $u_r = u_e$ and 0 otherwise. Matching the main with the inner layer and the external flow raises the BCs $F(0) = S(0) = 0$ and $F'(1) = 0$ $(I = 0)$ or $F'(1) = 1$ $(I = 1)$, $S(1) = 0$, respectively; the first can also be accepted from a pure kinematic point of view, cf. Eq. (3). These and a suitable model for l, see Eq. (49), supplement Eq. (53) to a parabolic eigenvalue problem with eigenvalues $\Delta(x)$ in terms of a non-trivial (vorticity-affected) solution $F' \not\equiv 1$, $S \not\equiv 0$.

The specification of the arising problem governing the flow in the main layer for $I = 1$ is straightforward. Free flows are self-preserving $(\partial_x F' \equiv \partial_x S \equiv 0)$ if $u_r(x)$ satisfies a power law $u_r \propto x^m$ with some constant exponent m, giving $\Delta \propto dx$ with $d\,(> 0)$ being an eigenvalue. This holds in any case for $I = 0$ due to matching the flow quantities with those of the transitional flow as $x \to 0_+$. Here, the most interesting case is the free (here planar) jet, cf. Fig. 1(b), described by the classical Schlichting problem[3]: $mF'^2 - (m+1)FF'' = S'/d$ subject to the homogeneous BCs, integration between $\eta = 0$ and 1 gives $m = -1/2$ for the second eigenvalue m governing the centre-line speed u_r and expressing conservation of axial momentum. As the solution F does not induce an external flow, higher order terms in Eq. (52) originate in the flow history for $x \ll 1$. The validity of its self-similar leading order structure for $x \gg 1$ and the associated variation of the momentum by entrainment are still under debate.

4.2. *Turbulent boundary layers: some exciting novelties*

At first, the asymptotic structure of (initially attached) turbulent BLs allows for a powerful categorisation in terms of the magnitude of the streamwise velocity defect related to the imposed surface speed $u_e(x)$ in the fully turbulent region[14]: $\Delta_u := 1 - \widetilde{u}(x,y)/u_e(x)$. Secondly, distinguishing between *slightly underdeveloped* and fully developed turbulence yields substantial analytical progress when it comes to the challenging problems of turbulent marginal and massive separation.

4.2.1. *Classification*

Let the fundamental velocity scale $\widetilde{u}_\tau$ serve as a reference scale. Then the following statements about the internal BL structure can be made.

Observation 2 (BL scaling). In the limit Re $\to \infty$, each sub-layer in a turbulent BL is characterised by an intrinsic velocity scale, u_t with $u_t \to 0$, that governs the turbulent dynamics such that $\sigma_{u_i''} = O(u_t)$.

This follows also intuitively from Eq. (14) and Proposition 1. Hence, u_t is also a local measure for Δ_u *and* u_t^2 for $\bar{\tau}$. Consider a sublayer of wall-normal extent δ_s, say. Since the shear-layer approximation of Eq. (16) gives $\bar{\rho}\widetilde{u}\,\partial_x\widetilde{u} + \cdots \sim \partial_y\bar{\tau}$, we have $u_e u_t/x \sim u_t^2/\delta_y$ by order-of-magnitude analysis. For x properly scaled such that $u_e = O(1)$, this implies $\delta_s = O(u_t)$ also and a time-mean vorticity of $O(i)$ as this is approximated by $u_e \partial_y \Delta_u$. Reverting to Eq. (33), we find the total shear stress as nearly constant across the near-wall sublayer by matching it with $\overline{\tau}$ in the small-deficit layer located on top of the latter. Hence, u_t is identified with $\widetilde{u}_\tau$ there, and matching $y\, \partial_y \widetilde{u}$ confirms the famous, widely accepted logarithmic law of the wall: $u^+ \sim \kappa^{-1} \ln y+ + O(1)$ $(y^+ \to \infty)$; κ is the von Kármán constant.

In turn, the well-known skin-friction law $\widetilde{u}_\tau \sim \kappa / \ln \mathrm{Re}$ as $\mathrm{Re} \to \infty$ arises from matching $\widetilde{u}$, so that the sublayer scalings, Eq. (33), are completed by

$$\overline{\delta} = O\left(\frac{1}{\ln \mathrm{Re}}\right), \quad \overline{\delta}_\nu = O\left(\frac{\ln \mathrm{Re}}{\mathrm{Re}}\right). \tag{54}$$

This finally gives the smallest scales determined by $\widetilde{u}_\tau$ and $\overline{\delta}_\nu$.[15] One then finds that those coincide with the celebrated Kolmogorov scales, whereas these cannot be identified in the fully turbulent regime. (They merely follow from Kolmogorov's first similarity hypothesis.[17])

These findings allow for the first classification of turbulent BLs:

(I) *Classical two-tiered*: $\Delta_u = O(\widetilde{u}_\tau)$, single velocity scale $\widetilde{u}_\tau$;

(II) *Three-tiered*: $\Delta_u = O(u_t)$, second turbulent velocity scale $u_t := \widetilde{u}_\tau^{2/3}$;

(III) *Four-tiered*: $\Delta_u \ll 1$, second velocity scale u_t of $O(\Delta_u)$;

(IV) *Four-tiered*: $\Delta_u = O(1)$, further velocity scales u_t of $O(1)$.

The situations (I) and (II) are both associated with the so-called small-deficit BLs and allow for a rigorous asymptotic analysis in the limit $\mathrm{Re} \to \infty$. The main-layer equations follow from expanding Eq. (53) with $s = O(\Delta_u^2)$.

The occurrence of an $O(1)$-velocity deficit as proposed by (IV) can only be understood if the slenderness of the BL is also measured by a slenderness parameter α taken as small, as for free shear flows. The rationale of Sec. 4.1 applies also here, and a justification is again by experimental evidence and the closure constants, cf. Eq. (49). But then the structure of a wake with $\Delta_u = O(1)$ applies to such a BL as far as the largest part of the fully turbulent regime is concerned. Hence, this wake region is two-tiered and gives rise to a smaller velocity scale of $O(\alpha^{3/4})$. Here, an appealing physical interpretation is possible: the presence of the wall is felt by the largest eddies in that wake region, increasingly with shrinking of the y-scale but just down to their diameter of $O(\alpha^{3/2})$ where their motion is blocked. The further near-wall region then exhibits a small-defect structure as applying in case (I): $\tilde{u}_\tau$ is the typical velocity scale but the deficit now around the $O(1)$-surface slip exerted by the wake flow. Thus, the BL is four-tiered: inner and outer layers, defect layer, viscous sublayer.

The "intermediate" category (III) then provides the missing link between (II) and (IV): in (IV), a genuine two-perturbation analysis based on α and Re is adopted following (III), where we the first time refrain from considering the BL scaling strictly in the limit Re $\rightarrow \infty$ but u_t just as small.

The level of turbulence intensity in a shear layer is advantageously epitomised by the so-called turbulence level parameter T_u,[3] here specified as an actual reference value of $\overline{K}$ related to that which would apply if turbulence was fully developed. One then characterises the BL flow as

(a) *Fully developed*: $T_u = 1$ (as considered so far);
(b) *Slightly underdeveloped*: $T_u \rightarrow 0$ as Re $\rightarrow \infty$.

This concept put forward by Neish and Smith[23] is of paramount importance when it comes to breakaway separation of the BL.

4.2.2. *Addressing turbulent separation: where are we now?*

There are two types of mean-flow separation from a perfectly smooth surface that can be described in a self-consistent manner. At first, a suitably controlled smooth adverse pressure gradient imposed on the BL by a fully attached external flow leads to so-called marginal or BL-internal separation, characterised by closed zones of reverse flow. Secondly, gross or breakaway separation means massively separated flow. As a well-known result of potential flow-theory, here the imposed pressure varies proportionally to $(x_s - x)^{1/2}$ immediately upstream of the separation point, $(x, y) = (x_s, 0)$, and regularly downstream of it.

Small-deficit BLs are fully insensitive to smooth adverse pressure gradients in terms of their tendency to separate. That is, they remain firmly attached as their fully turbulent main portion is in fact a predominantly inviscid, irrotational one. Hence, a description of that type of separation, which more precisely means states of marginal or BL-internal separation, has to consider an initially attached four-tiered large-deficit BL.[14] However, underdeveloped small-deficit BLs appear quite naturally as a result of laminar–turbulent transition close to a stagnation point of the external potential flow on a surface More precisely, the local viscous-inviscid interaction process regularising the above singularity at $x = x_s$ fixes the dependence $T_u(\mathrm{Re})$ as $\bar{\delta}_\nu/\bar{\delta}$ must vary predominantly algebraically with Re rather than exponentially as in Eq. (54). We are thus led to the following possibilities:

(A) $T_u \ll 1$: case (I) above, applies to massive separation *only*;
(B) $T_u = O(1)$: case (IV) above, applies to marginal separation.

Concerning (A), the line of research is initiated by Neish and Smith[23] and its status quo covered by Scheichl.[21]

Let us complete this survey by focussing on the largely unappreciated case (B) applied to massive separation. This proves physically attractive since it predicts fully developed turbulence already upstream of separation, expected to take place (sufficiently far) downstream of it. Scenario (A), however, then has to imply a kind of *secondary* or "turbulent-turbulent" transition towards this ultimate state, but such a mechanism has not been detected so far theoretically in the high-Re limit, neither for attached nor separated BLs. Furthermore, the Re-independent eddy viscosity in case (B) predicts separated shear layers belonging to the class of massive ones addressed in Sec. 4.1 but with the difference that these separate the free-stream flow from the weakly reversing/recirculating flow in the open/closed eddy emerging due to large-scale separation. They are pressure-driven and "carry the frozen state" of their near-wall structure at separation near their boundary with the latter flow region. Self-consistent ideas on the structure of the flow past an obstacle on the global scale, for both cases (A) and (B), are preliminary yet.

In striking contrast to laminar BLs, where ICs are well-defined, e.g., by the existence of a stagnation point and give rise to a well-posed parabolic problem, the situation for turbulent ones is more awkward as correct ICs have to be found by matching with the time-mean transitional flow. For the (2D) flow past an obstacle, the region of transition shrinks to a point, coinciding with the front stagnation point $x = y = 0$, say, as Re $\to \infty$. Hence, first the singular structure of Eq. (53) for $x \to 0_+$ has to be elucidated, implied by $u_e \sim cx + O(x^2)$ in this limit with some constant $c > 0$. We accordingly expand $[F, S] \sim [F_0, S_0](\eta) - [\gamma_F(x)F_1(\eta), \gamma_S(x)S_1(\eta)] + \cdots$ with $(\gamma_F, \gamma_S) \to (0_+, 0_+)$. Providing a flavour of the method of balancing at first largely unknown gauge functions (γ_1, Δ), as typically adopted in such situations, on the basis of this particular important case proves quite instructive.

Order-of-magnitude analysis of Eq. (53) shows that its left-hand balances its right-hand side in leading order if $\Delta \sim dx$ $(x \to 0_+)$:

$$F_0'^2 - 1 - 2F_0 F_0'' = \frac{S_0'}{d}, \quad F_0(0) = S_0(0) = 0, \quad F_0'(1) - 1 = S_0(1) = 0. \tag{55}$$

We additionally have in mind that $S_0 = l^2 F_0''|F_0''|$ and require forward flow: $F_0' > 0$ $(F_0 \geq 0)$. There are three scenarios: $F'(0) > 1$, $F'(0) < 1$, $F'(0) = 1$. The first/second refers to an overshooting-jet-/wake-type velocity profile with $S > 0/S < 0$ and $F'' > 0/F'' < 0$ for sufficiently small values of η. This means at least one local maximum/minimum $F_0'(\eta^*) >/< 1$ for some $\eta^* \in]0, 1[$ where F_0'', S_0 change sign. But then S_0 exhibits a local maximum/minimum for some $0 < \eta < \eta^*$, implying $S'(\eta^*) </> 0$, contradicting Eq. (55) (supported by attempts to solve this problem numerically).

We hence have to accept only the third possibility of the trivial (potential-flow) solution $F_0 \equiv \eta$, $S_0 \equiv 0$, associated with a velocity deficit of $O(\gamma_F)$ and $\gamma_S = \gamma_F^2$. By integration with respect to η, one finally verifies

$$c\eta F_1' - [c + \Delta (x\gamma_1)^{-2} d_x(x^2\gamma_1)]F_1 \sim S_1, \quad c := \lim_{x \to 0_+} (x\gamma_1)^{-1} d_x(x\Delta) \tag{56}$$

and $F_1 = S_1 = 0$ $(\eta = 0)$, $F_1' = S_1 = 0$ $(\eta = 1)$. Therefore, any nontrivial form of F_1, S_1 requires both c to be positive and the bracketed term in Eq. (56) to vanish. However, since the second added therein is non-negative, we again face a contradiction: the problem posed by Eq. (53) cannot be solved with ICs provided for arbitrarily small values of x. Hence, current activities focus on spontaneous secondary transition by a loss of parabolicity of the small-deficit equations referring to case (A).

5. Some Exercises

(1) Derive Eq. (16) from Eqs. (1) and (2).
(2) Derive Eqs. (25) and (31) from Eqs. (17) and (18).
(3) Outline the main ideas underlying the Re-independent time-mean scalings of turbulent shear layers having a large velocity deficit.
(4) Large-deficit BL immediately upstream of separation: show that the surface slip $u_e(x)F'(x,1)$ vanishes like $(-X)^{1/4}$ as $X := x - x_s \to 0_-$ for $u_e(x_s) > 0$ and $u_e(x) - u_e(x_s)$ vanishing like $(-X)^{1/2}$. Hint: consider the essential sublayer where $Y = O[(-X)^{1/3}]$ and all terms of Eq. (53) are retained, then show by matching that the solution proceeds downstream uniquely (*advanced*).
(5) Show that the initial-value problem for the BL posed by Eq. (53), $[F,T](x,0) = [0,0]$, $F'(x,1) - 1 = T(x,1) = 0$ and appropriate initial conditions for F, T, Δ prescribed for $x = 0$ is well-posed (strictly parabolic). Hint: here the essential sublayer is given by $Y = O(x^{1/3})$ as $x \to 0_+$ (*advanced*).

References

1. H. Tennekes and J. L. Lumley, *A First Course in Turbulence*. MIT Press, Cambridge, MA, 1972.
2. A. A. Townsend, *The structure of turbulent shear flows*, 2nd edn. Cambridge University Press 1976.
3. H. Schlichting and K. Gersten, *Boundary-Layer Theory*, 8th revised/enlarged ed. (corrected reprint, more recent revised 9th and 10th eds. available in German), Parts III–V. Springer, Berlin, Heidelberg, New York (2003). Specifically note Parts III–V.
4. S. B. Pope, *Turbulent Flows*. Cambridge University Press 2000. Corrections available for free download at http://pope.mae.cornell.edu.
5. M. Lesieur, *Turbulence in Fluids*, 4th edn. Springer, Dordrecht (2008).
6. P. A. Davidson, Y. Kaneda, K. Moffatt and K. Sreenivasan (eds), *A Voyage Through Turbulence*. Cambridge University Press, Cambridge, 2011. Book doi: 10.1017/CBO9781139018241.

7. R. A. W. M. Henkes, *Overview of Turbulence Models for External Aerodynamics* (Series 01: Aerodynamics 13). Delft Univesity Press (1998).

8. B. E. Launder and N. D. Sandham, *Closure Strategies for Turbulent and Transitional Flows.* Cambridge University Press, Cambridge, 2002. Book doi: 10.1017/CBO9780511755385.

9. D. C. Wilcox, *Turbulence Modelling for CFD*, 3rd edn., includes CD-ROM containing numerical software with user's guide. DCW Industries, Inc. (2006).

10. U. Frisch, *Turbulence: The Legacy of A. N. Kolmogorov.* Cambridge University Press (1995).

11. Prasad, K. Sreenivasan, Measurement and interpretation of fractal dimension of the scalar interface in turbulent flows, *Phys. Fluids A* **2**(5), 792–807 (1990).

12. M. Van Dyke, *An Album of Fluid Motion*, p. 56, Fig. 94. The Parabolic Press, Stanford (2002).

13. F. T. Smith, D. J. Doorly and A. P. Rothmayer, On displacement-thickness, wall-layer and mid-flow scales in turbulent boundary layers, and slugs of vorticity in channel and pipe flows, *Proc. R. Soc. Lond. A* **428**(1875), 255–281 (1990).

14. A. Kluwick and B. Scheichl, High-reynolds-number asymptotics of turbulent boundary layers: From fully attached to marginally separated flows. In *BAIL 2008 — Boundary and Interior Layers* (Lecture Notes in Computational Science and Engineering, Vol. 69), eds A. F. Hegarty *et al.*, pp. 3–22. Springer, Berlin, Heidelberg (2009).

15. C. R. Smith, J. D. A. Walker, A. H. Haidari and U. Sobrun, On the dynamics of near-wall turbulence (Turbulent flow wtructure near walls. Part II), *Phil. Trans. R. Soc. Lond. A* **336**(1641), 131–175 (1991).

16. R. S. Rivlin and J. L. Ericksen, Stress-deformation relations for isotropic materials, *J. Rational Mech. Anal.* **4**, 323–425 (1955).

17. A. N. Kolmogorov, A refinement of previous hypotheses concerning the local structure of turbulence in a viscous incompressible fluid at high Reynolds number, *J. Fluid Mech.* **13**(1), 82–85 (1962).

18. L. Prandtl, *The Role of Viscosity in the Mechanism of Developed Turbulence*, GOAR 3712, DLR Archive (1945).

19. E. Deriat and J. P. Guiraud, On the asymptotic description of turbulent boundary layers, *J. Theor. Appl. Mech. (J. Méc. Théor. Appl.)*, Special Issue: Asymptotic Modelling of Fluid Flows (Numéro spécial: Modelisation Asymptotique D'Ecoulements de Fluides), 109–140 (1986).

20. S. Cowley, Laminar Boundary Layer Theory: A 20th Century Paradox?. In *Mechanics for a New Millenium. Proc. ICTAM 2000*, eds H. Aref and J. W. Phillips, pp. 389–412. Kluwer, Dordrecht (2002). Long version: http://www.damtp.cam.ac.uk/user/sjc1/papers/ictam2000/long.pdf.

21. In: J.-S.-B. Gajjar (ed.) *Phil. Trans. R. Soc. A* **372**(2020), Theme Issue: Stability, separation and close-body interactions (2014).

22. W. Schneider, Boundary-Layer theory of free turbulent shear flows, *Z. Flugwiss. Weltraumforsch. (J. Flight Sci. Space Res.)* **15**(3), 143–158 (1991).

23. A. Neish and F. T. Smith, On turbulent separation in the flow past a bluff body. *J. Fluid Mech.* **241**, 443–467 (1992).

Chapter 4

Nonlinear Free Surface Flows with Gravity and Surface Tension

J.-M. Vanden-Broeck

Department of Mathematics, University College London,
Gower Street, London, UK
j.vanden-broeck@ucl.ac.uk

This chapter is concerned with the computation of nonlinear free surface flows. Both the effects of surface tension and gravity are included in the dynamic boundary condition. Special attention is devoted to the singular behaviour at the points where free surfaces intersect rigid walls. Applications to bubbles rising in a fluid, flows emerging from a nozzle and cavitating flows are presented. It is shown how physical solutions are selected in the limit as the surface tension tends to zero.

1. Introduction

Free surface problems occur in many aspects of science and everyday life. They can be defined as problems whose mathematical formulation involves surfaces that have to be found as part of the solution. Such surfaces are called free surfaces. Examples of free surface problems are waves on a beach, bubbles rising in a glass of champagne, melting ice, flows pouring from a container and sails blowing in the wind. In these examples, the free surface is the surface of the sea, the interface between the gas and the champagne, the surface of the ice, the boundary of the pouring flow and the surface of the sail.

In this chapter, we concentrate on applications arising in fluid mechanics. We restrict our attention to steady, inviscid, irrotational and two-dimensional (2D) flows. The effects of gravity and surface tension are included in the nonlinear dynamic boundary condition.

Free surface flows fall into two main classes. The first is the class of such flows for which there are intersections between the free surface and a rigid surface. The classic example in this class is the flow due to a ship moving at the surface of a lake, which involves an intersection between the free surface and a rigid surface (i.e., the hull of the ship). Other examples are jets leaving a nozzle, bubbles attached to a wall and flows under a sluice gate. In each case there is a rigid surface (the nozzle, the obstacle, the wall or the gate) that intersects a free surface. The second class contains free surface flows for which there are no intersections between the free surface and a rigid wall. Here the classic example is the flow due to a submerged object moving below the surface of a lake. Other examples include free bubbles rising in a fluid and solitary waves. This chapter is concerned with the theory of free surface flows of the first class. We proceed in stages of increasing complexity.

Due to space limitation, some of the details are omitted. The reader is referred to the monograph "Gravity-Capillary Free Surface Flows" (see Ref. 5) for a complete presentation, more examples and more references. Some of the missing parts are also suggested as exercises at the end of the chapter.

2. Basic Concepts

We first review some equations of fluid mechanics which will be used in this chapter. For further details, see, for example, Refs. 1 and 2. All the fluids considered are assumed to be inviscid and to have constant density ρ (i.e., to be incompressible).

Conservation of momentum yields the Euler's equations

$$\frac{D\mathbf{u}}{Dt} = -\frac{1}{\rho}\nabla p + \mathbf{X}, \tag{1}$$

where $\mathbf{u}$ is the vector velocity, p is the pressure and $\mathbf{X}$ is the body force. Here

$$\frac{D}{Dt} = \frac{\partial}{\partial t} + \mathbf{u} \cdot \nabla \tag{2}$$

is the material derivative. We assume that the body force $\mathbf{X}$ derives from a potential Ω, i.e.,

$$\mathbf{X} = -\nabla\Omega. \tag{3}$$

The flows are assumed to be irrotational. Therefore,

$$\nabla \times \mathbf{u} = 0. \tag{4}$$

Relation (4) implies that we can introduce a potential function ϕ such that

$$\mathbf{u} = \nabla\phi. \tag{5}$$

Conservation of mass gives

$$\nabla \cdot \mathbf{u} = 0. \tag{6}$$

Then (5) and (6) imply the Laplace equation

$$\nabla^2\phi = 0. \tag{7}$$

Flows which satisfy (4)–(7) are referred to as potential flows.

After integration, (1) gives (after some algebra) the well known Bernoulli equation

$$\frac{\partial\phi}{\partial t} + \frac{\mathbf{u}\cdot\mathbf{u}}{2} + \frac{p}{\rho} + \Omega = F(t). \tag{8}$$

Here, $F(t)$ is an arbitrary function of t. It can be absorbed in the definition of ϕ and (8) can be rewritten as

$$\frac{\partial\phi}{\partial t} + \frac{\mathbf{u}\cdot\mathbf{u}}{2} + \frac{p}{\rho} + \Omega = B, \tag{9}$$

where B is a constant. For steady flows, (9) reduces to

$$\frac{\mathbf{u}\cdot\mathbf{u}}{2} + \frac{p}{\rho} + \Omega = B. \tag{10}$$

3. Two-Dimensional Flows

Many interesting free surface flows can be modelled as 2D flows. We then introduce Cartesian coordinates x and y with the y-axis directed vertically upwards (we reserve the letter z to denote the complex quantity $x + iy$). In the applications considered in this chapter, the body potential in (8) is due to gravity. Assuming that the acceleration of gravity g is acting in the negative y-direction, we write Ω as

$$\Omega = gy. \tag{11}$$

An example of a 2D free surface is illustrated in Fig. 1. The fluid (e.g., water) is bounded below by the bottom and a circular obstacle. The flow is from left to right. The upper curve is the interface between the fluid and the atmosphere with is assumed to be characterised by a constant atmospheric pressure p_a. We refer to such an interface as a free surface. The 2D configuration of Fig. 1 provides a good approximation for the three-dimensional (3D) free surface flow past a long cylinder perpendicular to the plane of the figure (except near the ends of the cylinder). The cross-section of the cylinder is the half circle shown in Fig. 1.

For 2D potential flows, (4) and (6) become

$$\frac{\partial u}{\partial y} - \frac{\partial v}{\partial x} = 0 \tag{12}$$

and

$$\frac{\partial u}{\partial x} + \frac{\partial v}{\partial y} = 0. \tag{13}$$

Here, u and v are the x and y components of the velocity vector $\mathbf{u}$.

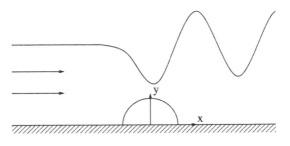

Fig. 1.   Sketch of the 2D free surface flow past a submerged circle.

We can introduce a streamfunction ψ by noting that (13) is satisfied by writing

$$u = \frac{\partial \psi}{\partial y}, \tag{14}$$

$$v = -\frac{\partial \psi}{\partial x}. \tag{15}$$

It then follows from (12) that

$$\nabla^2 \psi = \frac{\partial^2 \psi}{\partial x^2} + \frac{\partial^2 \psi}{\partial y^2} = 0. \tag{16}$$

For 2D flows, Eqs. (5) and (7) give

$$u = \frac{\partial \phi}{\partial x}, \tag{17}$$

$$v = \frac{\partial \phi}{\partial y} \tag{18}$$

and

$$\nabla^2 \phi = \frac{\partial^2 \phi}{\partial x^2} + \frac{\partial^2 \phi}{\partial y^2} = 0. \tag{19}$$

Combining (14), (15), (17) and (18) we obtain

$$\frac{\partial \phi}{\partial x} = \frac{\partial \psi}{\partial y}, \tag{20}$$

$$\frac{\partial \phi}{\partial y} = -\frac{\partial \psi}{\partial x}. \tag{21}$$

Equations (20) and (21) can be recognised as the classical Cauchy–Riemann equations. They imply that the complex potential

$$f = \phi + i\psi \tag{22}$$

is an analytic function of $z = x + iy$ in the flow domain. This result is particularly important since it implies that 2D potential flows can be investigated by using the theory of analytic functions. This applies, in particular, to all 2D potential free surface flows with or without gravity or surface tension included in the dynamic boundary condition. It does not, however, apply to axisymmetric and 3D free surface flows. Since the derivative of an analytic function is also an analytic function, it follows that the complex velocity

$$u - iv = \frac{\partial \phi}{\partial x} - i\frac{\partial \phi}{\partial y} = \frac{\partial \psi}{\partial y} + i\frac{\partial \psi}{\partial x} = \frac{df}{dz} \tag{23}$$

is also an analytic functions of $z = x + iy$. The theory of analytic functions will be used intensively in the following sections to study 2D free surface flows.

We now show that for steady flows the streamfunction ψ is constant along streamlines. A streamline is a line to which the velocity vectors are tangent. Let us describe a streamline in parametric form by $x = X(s)$, $y = Y(s)$, where s is the arclength. Then we have

$$-vX'(s) + uY'(s) = 0, \tag{24}$$

where the primes denote derivatives with respect to s. Using (14) and (15), we have

$$\frac{\partial \psi}{\partial x}X'(s) + \frac{\partial \psi}{\partial y}Y'(s) = \frac{d\psi}{ds} = 0 \tag{25}$$

which implies that ψ is a constant along a streamline. For steady flows the kinematic boundary condition implies that a free surface is a streamline. The streamfunction is then constant along a free surface.

An important challenge in finding solutions for flows like that of Fig. 1 is that the shape of the free surface is not known *a priori*: it has to be found as part of the solution. It is then necessary to impose an extra condition on the free surface. This is known as the dynamic boundary condition. It can be derived as follows. First, we introduce the concept of surface tension by writing

$$p - p_a = \frac{T}{K}. \tag{26}$$

Here, p denotes the pressure just below the free surface, T the coefficient of surface tension (assumed to be constant) and K the curvature of the free surface. The dynamic boundary condition is then obtained by substituting (11) and (26) into (9) evaluated just below the free surface. This yields

$$\frac{\partial \phi}{\partial t} + \frac{1}{2}(\phi_x^2 + \phi_y^2) + gy + \frac{T}{\rho}K = B. \tag{27}$$

If we denote by θ the angle between the tangent to the free surface and the horizontal, then the curvature K can be defined by

$$K = -\frac{d\theta}{ds}, \tag{28}$$

where s denotes again the arclength. In particular if the (unknown) equation of the free surface is $y = \eta(x, t)$, then

$$\tan \theta = \eta_x \quad \text{and} \quad \frac{dx}{ds} = \frac{1}{(1 + \eta_x^2)^{\frac{1}{2}}}. \tag{29}$$

Using (28), (29) and the chain rule gives the formula

$$K = -\frac{\eta_{xx}}{(1 + \eta_x^2)^{\frac{3}{2}}}. \tag{30}$$

4. The Mathematical Model

We shall study in detail the 2D free surface flow sketched in Fig. 2. The flow domain is bounded below by the horizontal wall AB and above by the inclined walls CD and DE and by the free surface EF. The fluid is assumed to be steady. We introduce Cartesian coordinates with the x-axis along the horizontal wall AB and the y-axis through the separation point E (here a separation point refers to an intersection between a free surface and a rigid wall). The angles between the walls CD and DE and the horizontal are denoted by γ_1 and γ_2 respectively.

We denote by μ the angle between the free surface and the wall at the point E. When $\mu = \pi$, the free surface is tangent to the wall at E. When $0 < \mu < \pi$, there is locally a flow inside an angle near E and the velocity at E is zero. When $\mu > \pi$, we have a flow around an angle near E and the velocity at E is infinite. We shall see in the next sections that these three cases can occur.

The configuration of Fig. 2 was chosen not only because it is simple but also because it can be used to describe many properties

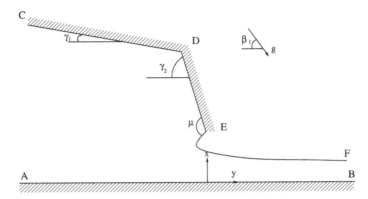

Fig. 2. A 2D free surface flow bounded by the walls CD, DE and AB and the free surface EF. The separation point E is defined as the point at which the free surface EF intersects the wall DE. The points C, A, F and B are at infinite distance from E. The flow is from left to right

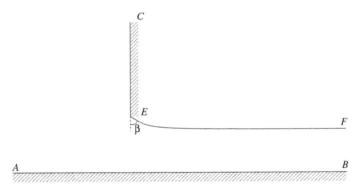

Fig. 3. Sketch of the free surface flow under a gate. The flow is from left to right

of free surface flows which intersect rigid walls. These properties when understood for the flow of Fig. 2 can then be used to describe locally flows with more complex geometries.

There are various interpretations of the flow of Fig. 2. The first is the flow emerging from a container bounded by the walls CD, DE and AB. When $\gamma_1 = \gamma_2 = \pi/2$, the configuration of Fig. 2 models the flow under a gate (see Fig. 3). Here, the point D is irrelevant and was omitted from the Fig. 3.

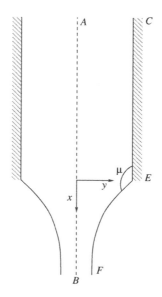

Fig. 4.   The free surface flow emerging from a nozzle

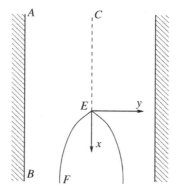

Fig. 5.   A "bubble" rising in a tube, viewed in a frame of reference moving with bubble. Physical bubbles are characterised by a continuous slope at the apex

Further particular cases of Fig. 2, which model bubbles rising in a fluid and jets falling from a nozzle are illustrated in Figs. 4 and 5.

As mentioned in the introduction, we will proceed with problems of increasing complexity. Section 5 is devoted to free surface flows with $g = 0$ and $T = 0$. Such flows are called free streamline flows and the corresponding free surfaces, free streamlines. In Sec. 6, we will study the effect of surface tension $(T \neq 0, g = 0)$. In Sec. 7, we will examine the effect of gravity $(T = 0, g \neq 0)$. The combined effects of gravity and surface tension $(T \neq 0, g \neq 0)$ are considered in Sec. 8.

5. Free Streamline Flows: $g = 0$, $T = 0$

5.1. *Forced separation*

We consider the flow configuration of Fig. 2. The effects of gravity and surface tension are neglected $(T = 0, g = 0)$. We refer to this problem as one of forced separation because the free surface is "forced" to separate at the point E where the wall DE terminates. Following the notations of Sec. 3, we introduce the complex potential function $f = \phi + i\psi$ and the complex velocity $u - iv$.

The wall AB is a streamline along which we choose $\psi = 0$. The walls CD and DE and the free surface EF define another streamline along which the constant value of ψ is denoted by Q. We also choose $\phi = 0$ at the separation point E. These two choices ($\psi = 0$ on AB and $\phi = 0$ at E) can be made without loss of generality because ϕ and ψ are defined up to arbitrary additive constants. Bernoulli's equation (10) with $\Omega = 0$ yields

$$\frac{1}{2}(u^2 + v^2) + \frac{p}{\rho} = \text{constant} \qquad (31)$$

everywhere in the fluid. The free surface EF separates the fluid from the atmosphere which is assumed to be characterised by a constant pressure p_a. In the abscence of surface tension, the pressure is continuous across the free surface (see (26)). Therefore, $p = p_a$ on the free surface. It follows from (31) that

$$u^2 + v^2 = U^2 \quad \text{on } EF, \qquad (32)$$

where U is a constant.

A significant simplification in the formulation of the problem is obtained by using ϕ and ψ as independent variables. This choice

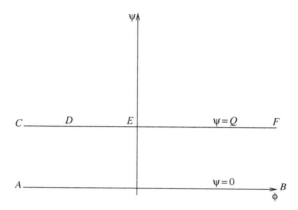

Fig. 6.   The flow configuration of Fig. 2 in the complex potential plane $f = \phi + i\psi$

was used before by many investigators (see Ref. 5). We shall use it extensively in our studies of gravity capillary free surface flows. The simplification comes from the fact that the flow domain is mapped into the strip $0 < \psi < Q$ shown in Fig. 6.

The free surface EF (whose position was unknown in the physical plane $z = x + iy$ of Fig. 2) is now part of the known boundary $\psi = Q$ in the $f = \phi + i\psi$-plane. Since $u - iv$ is an analytic function of z and z is an analytic function of f (the inverse of an analytic function is also an analytic function), $u - iv$ is an analytic function of f.

A remarkable result is that many free streanline problems can be solved in closed form (see Refs. 3 and 4). These exact solutions are obtained by using conformal mappings and several methods have been developed to calculate them. The method we chose to describe, uses a mapping of the flow domain into the unit circle. It was chosen because it yields naturally to the series truncation methods used in Secs. 6–8 to solve numerically problems with gravity and surface tension included.

In the absence of gravity and surface tension, the flow approaches a uniform stream of constant depth H as $x \to \infty$. It follows from the dynamic boundary condition (32) that this uniform stream is characterised by a constant velocity U. Since $\psi = 0$ on AB and $\psi = Q$ on EF, $H = Q/U$. We introduce dimensionless variables by using U as the reference velocity and H as the reference length.

Therefore, $\psi = 1$ on the walls CD and DE and on the free surface EF. The dynamic boundary condition (32) becomes

$$u^2 + v^2 = 1 \quad \text{on } EF. \tag{33}$$

We define the logarithmic hodograph variable $\tau - i\theta$ by the relation

$$u - iv = e^{\tau - i\theta}. \tag{34}$$

The function $\tau - i\theta$ has some interesting properties. Firstly, the quantity $\tau = \frac{1}{2}\ln(u^2 + v^2)$ is constant along free streamlines (see (32)). Secondly, θ can be interpreted as the angle between the vector velocity and the horizontal. Thirdly, (34) leads, for steady flows, to a very simple formula for the curvature of a streamline. This formula can be derived as follows. Since the vector velocity is tangent to streamlines, θ is the angle between the tangent to a streamline and the horizontal. The curvature K of a streamline is then given by (28). Using the chain rule, we rewrite (28) as

$$K = -\frac{\partial \theta}{\partial \phi}\frac{\partial \phi}{\partial s} - \frac{\partial \theta}{\partial \psi}\frac{\partial \psi}{\partial s}. \tag{35}$$

Along a streamline ψ is constant and therefore

$$\frac{\partial \psi}{\partial s} = 0 \quad \text{and} \quad \frac{\partial \phi}{\partial s} = e^{\tau}. \tag{36}$$

Substituting (36) into (35) yields the simple formula

$$K = -e^{\tau}\frac{\partial \theta}{\partial \phi}. \tag{37}$$

We map the strip of Fig. 6 into the unit circle in the t-plane by the conformal mapping

$$e^{-\pi f} = \frac{(1-t)^2}{4t}. \tag{38}$$

The flow configuration in the t-plane is shown in Fig. 7. It can easily be checked that the points A and C are mapped into $t = 0$ and that the points B and F are mapped into $t = 1$. The value of t

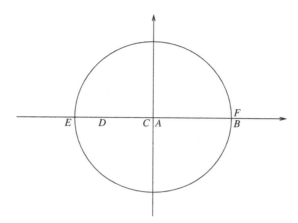

Fig. 7. The flow configuration of Fig. 6 in the complex t-plane. Here we sketch values of the imaginary value of t versus the real part of t

at point D is denoted by $t = d$. The free surface EF is mapped onto the portion

$$t = e^{i\sigma}, \quad 0 < \sigma < \pi \tag{39}$$

of the unit circle. This can easily be shown by noting that a substitution of (39) into (38) gives after some algebra

$$\phi = -\frac{1}{\pi} \ln \sin^2 \frac{\sigma}{2}, \quad \psi = 1. \tag{40}$$

As σ varies from 0 to π, ϕ varies from ∞ to 0, so that (39) is the image of the free surface in the t-plane.

One might attempt to represent the complex velocity $w = u - iv$ by the series

$$w = \sum_{n=0}^{\infty} a_n t^n. \tag{41}$$

However, the series will not converge inside the unit circle $|t| \leq 1$, because singularities can be expected at the corner D and as $x \to -\infty$ (i.e., at $t = 0$). We can, however, generalise the representation (41) by writing

$$w = G(t) \sum_{n=0}^{\infty} a_n t^n \tag{42}$$

where the function $G(t)$ contains all the singularities of w. As we shall see in Secs. 6–8 this type of series representation enables the accurate calculation of many free surface flows with gravity and surface tension included. For the present problem we require $G(t)$ to behave like w as $t \to 0$ and as $t \to d$. Here d is the value of t at the point D in Fig. 7. We can then expect the series in (42) to converge for $|t| \leq 1$.

To construct $G(t)$, we find the asymptotic behaviour of w near the singularities by performing local asymptotic analysis near D and as $x \to -\infty$.

The flow near D is a flow inside a corner. It can be shown that the general solution for a flow in a corner γ is

$$z \approx A f^{\frac{\gamma}{\pi}}, \tag{43}$$

where A is a constant. It follows from (43) that

$$w \approx \frac{\pi}{A\gamma} f^{1-\frac{\gamma}{\pi}}. \tag{44}$$

When $\gamma < \pi$, the flow is inside at the angle and (44) implies that the velocity at the apex is zero. When $\gamma > \pi$, the flow is around the angle and (44) implies that the velocity at the apex is infinite.

For the flow of Fig. 2, $\gamma = \pi - \gamma_2 + \gamma_1$ and (44) imply

$$w = (f - \phi_D - i)^{\left(\frac{\gamma_2 - \gamma_1}{\pi}\right)} \quad \text{as } f \to \phi_D + i \tag{45}$$

where ϕ_D is the value of ϕ at the point D. Using (38), yields

$$f - \phi_D - i \approx t - d \quad \text{as} \quad f \to \phi_D + i. \tag{46}$$

Combining (45) and (46) gives

$$w \approx (t - d)^{\left(\frac{\gamma_2 - \gamma_1}{\pi}\right)} \quad \text{as } t \to d. \tag{47}$$

This concludes our local analysis near the point D.

As $x \to -\infty$, the flow behaves like the flow due to a sink at $x = y = 0$. Therefore,

$$f \approx -B \ln z \quad \text{as} \quad x \to -\infty, \tag{48}$$

where B is a positive constant. Differentiating (48) with respect to z gives

$$w = \frac{df}{dz} = -\frac{B}{z}. \tag{49}$$

Since the flux of the fluid coming from $-\infty$ is 1 and the angle between the walls CD and AB is γ_1, we have

$$B = \frac{1}{\gamma_1}. \tag{50}$$

Eliminating z between (48) and (49) gives

$$w = O[e^{\gamma_1 f}] \quad \text{as } f \to -\infty. \tag{51}$$

Relation (38) implies

$$e^{\pi f} = O(t) \quad \text{as} \quad f \to -\infty. \tag{52}$$

Therefore, (51) and (52) give

$$w = O(t^{\frac{\gamma_1}{\pi}}) \quad \text{as} \quad t \to 0. \tag{53}$$

Combining (47) and (53), we can choose

$$G(t) = (t - d)^{\frac{(\gamma_2 - \gamma_1)}{\pi}} t^{\frac{\gamma_1}{\pi}} \tag{54}$$

and write (42) as

$$w = (t - d)^{(\gamma_2 - \gamma_1)/\pi} t^{\gamma_1/\pi} \sum_{n=0}^{\infty} a_n t^n. \tag{55}$$

There are of course many other possible choices for $G(t)$. For example, $G(t)$ can be multiplied by any function analytic in $|t| \leq 1$.

We now need to determine the coefficients a_n in (55) so that the dynamic boundary condition (33) is satisfied. This can be done numerically by truncating the infinite series in (55) after N terms and finding the coefficients a_n, $n = 0 \ldots, N-1$ by collocation. This is the approach we will use when solving problems with the effects of gravity or surface tension included in the dynamic boundary condition. However, it can checked that the problem has the exact solution

$$w = \left[\frac{t - d}{1 - td}\right]^{\frac{(\gamma_2 - \gamma_1)}{\pi}} t^{\frac{\gamma_1}{\pi}}. \tag{56}$$

It then follows from (55) that

$$\sum_{n=1}^{\infty} a_n t^n = \left[\frac{1}{1 - td} \right]^{\frac{(\gamma_2 - \gamma_1)}{\pi}}. \tag{57}$$

The existence of an exact solution follows from the general theory of free streamline flows. This theory was developed by Kirchhoff and Helmoltz (see Refs. 3 and 4 for details).

The free surface profile is obtained by setting $\psi = 1$ in (56), calculating x_ϕ and y_ϕ from the identity

$$x_\phi + iy_\phi = \frac{1}{w} \tag{58}$$

and integrating with respect to ϕ.

As an example, let us assume that $\gamma_1 = \gamma_2 = \pi/2$ (see Fig. 3). Then (56) reduces to

$$w = t^{\frac{1}{2}} \tag{59}$$

and (39), (58) and (59) yield

$$x_\phi + iy_\phi = e^{-\frac{i\sigma}{2}} \quad \text{on} \quad 0 < \sigma < \pi \tag{60}$$

on the free surface EF. Differentiating (40) with respect to σ and applying the chain rule to (60) gives

$$x_\sigma + iy_\sigma = -\frac{1}{\pi} \cotan \frac{\sigma}{2} e^{-\frac{i\sigma}{2}}. \tag{61}$$

Integrating (61) gives the free profile in parametric form. It is shown in Fig. 8.

5.2.   *Free separation*

In Figs. 2 and 3, the free surface is forced to separate from the rigid wall DE at E because the wall DE terminates at E. We refer to this situation as forced separation. On the other hand, if the wall DE is replaced by a smooth curve then the point of separation E can be in principle any point on the smooth curve. We refer to this situation as free separation.

A typical example of free separation is the cavitating flow past a circle (see Fig. 9).

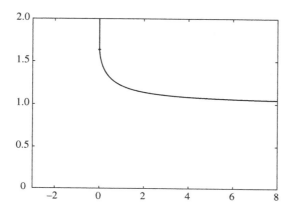

Fig. 8. Computed free surface profile (values of y versus x) for the flow configuration of Fig. 3. The position of the separation point E is indicated by a small horizontal line. The vertical scale has been exaggerated to show clearly the free surface profile. The bottom is on $y = 0$

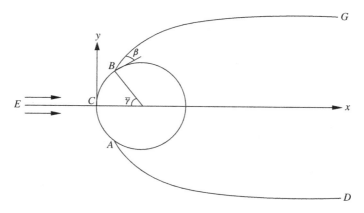

Fig. 9. The cavitating flow past a circle in an unbounded fluid domain. When the surface tension T is equal to zero the free surfaces leave the circle tangentially and $\beta = 0$. When $T \neq 0$ the angle β can be different from zero

The cavity is bounded by the two free surfaces BG and AD. It is characterised by a constant pressure p_a and it is open as $x \to \infty$. The position of the separation points A and B is characterised by the angle $\bar{\gamma}$. The angle between the free surfaces and the circle at the separation points B and A is denoted by β. Since we assumed in this section that $g = 0$ and $T = 0$, we have $\beta = 0$. We will see in the next sections that β can be different from zero when $T \neq 0$. There no exact solutions for the flow of Fig. 9 because the rigid boundaries are not polygonal. However, solutions can be calculated numerically by series truncation. Details can be found in Vanden-Broeck.[5] The results show that the value of the angle $\bar{\gamma}$ does not come as part of the solution. In other words, there is a flow for each value of $\bar{\gamma}$. A natural question is: Which value of $\bar{\gamma}$ will be selected in an experiment? One way to select a solution is to impose an extra condition known as the Brioullin condition (see Refs. 3 and 4 for details). It leads to the values

$$\gamma^* \approx 55^0. \tag{62}$$

As we shall see in Sec. 6.2, an alternative way to achieve this selection is to introduce the surface tension T in the problem and then to take the limit $T \to 0$.

6. Pure Capillary Free Surface Flows: $g = 0$, $T \neq 0$

6.1. *Forced separation*

In this section, we will investigate the effects of the surface tension T on the free streamline solutions of Sec. 5. We show that the limit $T \to 0$ is singular. When $T \neq 0$, discontinuities can appear at the separation points. In particular, values of $\mu \neq \pi$ and $\beta \neq 0$ can occur in Figs. 2 and 3.

We can calculate nonlinear solutions for the flow configuration of Fig. 2 by modifying appropriately the series representation (55) to accomodate the singularity at $t = -1$. The flow near $t = -1$ is a flow in an angle μ. Using (44), we obtain

$$w \sim f^{1-\frac{\mu}{\pi}} \quad \text{as } \phi \to 0. \tag{63}$$

Using (38), we have

$$w \sim (t+1)^{2-\frac{2\mu}{\pi}} \quad \text{as} \quad t \to -1. \tag{64}$$

Therefore,

$$w = (t-d)^{\frac{(\gamma_2-\gamma_1)}{\pi}} t^{\frac{\gamma_1}{\pi}} (t+1)^{2-\frac{2\mu}{\pi}} \sum_{n=0}^{\infty} a_n t^n \tag{65}$$

is the appropriate generalisation of (55) when surface tension is included.

We present explicit calculations in the particular case $\gamma_1 = \gamma_2 = \pi/2$. In other words we consider the flow configuration of Fig. 3. The expression (65) becomes

$$w = t^{\frac{1}{2}}(t+1)^{-\frac{2\beta}{\pi}} \sum_{n=0}^{\infty} a_n t^n, \tag{66}$$

where β is defined in Fig. 3.

The dynamic boundary condition is given in dimensionless variables $(U = 1, H = 1)$ by

$$\frac{1}{2}(u^2 + v^2) + \frac{2}{\alpha_v} K = \text{constant}, \tag{67}$$

where α_v is defined by

$$\alpha_v = \frac{2\rho U^2 H}{T}. \tag{68}$$

Since $u^2 + v^2 \to 1$ and $K \to 0$ as $\phi \to \infty$, the constant on the right-hand side of (67) is equal to $1/2$.

We truncate the infinite series in (66) after N terms. We calculate the coefficients a_n, $n = 0, \ldots, N-1$ and β by satisfying (67) (with K rewritten in terms of u and v by using (37)) at the $N+1$ equally spaced mesh points

$$\sigma_I = \frac{\pi}{N+1}\left(I - \frac{1}{2}\right) \quad I = 1, \ldots, N+1. \tag{69}$$

This leads to a system of $N+1$ equations with $N+1$ unknowns which is solved by Newton's method.

Typical free surface profiles are shown in Fig. 10. For $\alpha_v = \infty$, the free surface profile reduces to the free streamline solution of Fig. 8. As $\alpha_v \to 0$, the free surface profile approaches the horizontal line $y = 1$. This is consistent with the fact that the dynamic boundary condition (67) predicts that the curvature of the free surface tends to zero as $\alpha_v \to 0$ (the line $y = 1$ has zero curvature).

Numerical values of β versus α_v are shown in Fig. 11.

As α_v varies from 0 to ∞, β varies continuously from $\pi/2$ to 0.

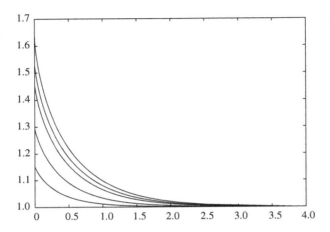

Fig. 10. Computed free surface profiles (values of y versus x) for the flow configuration of Fig. 3. The profiles from top to bottom correspond to $\alpha_v = \infty$, $\alpha_v = 50$, $\alpha_v = 25$, $\alpha_v = 10$ and $\alpha_v = 5$. The bottom is on $y = 0$.

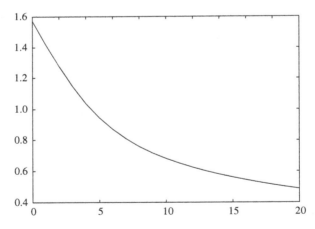

Fig. 11. Values of the angle β between the free surface and the wall at the separation point E (see Fig. 3) versus α_v.

6.2. *Free separation*

We now consider the open cavity model of Fig. 9 with the effect of the surface tension T included in the dynamic boundary condition. The results presented in the previous section suggests that the angle

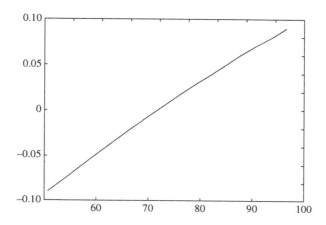

Fig. 12.   Values of β/π versus $\bar{\gamma}$ for $\bar{\alpha} = 1$

β in Fig. 9 will be different from zero when $T \neq 0$. This is confirmed by solving the problem numerically by using the series truncation procedure outlined in Sec. 5.1. The reader is referred to Ref. 5 for details. We present in Fig. 12 values β/π versus $\bar{\gamma}$ for $\bar{\alpha} = 1$. Here, $\bar{\alpha}$ is defined by

$$\bar{\alpha} = \frac{\rho U^2 R}{T}, \tag{70}$$

where R is the radius of the circle.

Figure 12 illustrates the fact that for each value $\bar{\alpha}$ (i.e., for each value of T) there is only one value of $\bar{\gamma}$ for which $\beta = 0$. We describe these particular values of $\bar{\gamma}$ by the function $\bar{\gamma} = g(\bar{\alpha})$. This means that for $\bar{\gamma} = g(\bar{\alpha})$ the free surface leaves the circle tangentially. The numerical calculations show that

$$g(\bar{\alpha}) \to \gamma^* \quad \alpha \to \infty, \tag{71}$$

where γ^* is defined in (62).

Therefore, a unique solution is obtained in the limit as T tends to zero. This shows that the solution satisfying the Brioullin condition can be selected by including surface tension in the problem and then taking the limit $T \to 0$.

7.   Pure Gravity Free Surface Flows: $g \neq 0$, $T = 0$

For pure gravity flows, the angles μ and β in Figs. 2 and 3 come also as part of the solution. However, they are restricted to a few values. For example, for the flow of Fig. 4, μ can only take one of the three values $\pi/2$, $2\pi/3$ and π. Here (and in the remaining part of this chapter) we assume that gravity is acting vertically downwards. Numerical solutions can be obtained by adapting appropriately the series truncation method of Secs. 5 and 6. Details can be found in Ref. 5.

We present results for the flow sketched in Figs. 4 and 5. We first note that these two flows are equivalent by symmetry: the free surfaces EF in Figs. 4 and 5 are identical. We chose to describe the results by refering to Fig. 5. Following the notations in Ref. 5, we introduce the Froude number

$$F = \frac{U}{\sqrt{gh}} \tag{72}$$

and the parameter

$$\nu = 2\frac{\pi - \mu}{\pi}, \tag{73}$$

where h is the distance between the two vertical walls in Fig. 5.

Figure 13 shows numerical values of ν versus F. These results imply

$$\nu = 1, \quad \mu = \frac{\pi}{2} \quad \text{for } 0 < F < F_c \tag{74}$$

$$\nu = \frac{2}{3}, \quad \mu = \frac{2\pi}{3} \quad \text{for } F = F_c \tag{75}$$

$$\nu = 0, \quad \mu = \pi \quad \text{for } F > F_c, \tag{76}$$

where

$$F_c \approx 0.3578. \tag{77}$$

Physically relevant bubbles should have a continuous slope at their apex. This occurs for all values of $0 < F < F_c$. Experiments show that bubbles are only observed for

$$F_e \approx 0.25. \tag{78}$$

This value is clearly in the interval $0 < F < F_c$. However, we do not have at this stage any criterion to select this particular solution. The

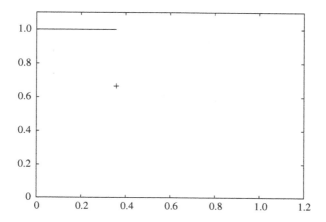

Fig. 13.   Values of ν versus F when $T = 0$.

selection will again be achieved in the next section by introducing the surface tension T and taking the limit $T \to 0$.

8.   Gravity-Capillary Free Surface Flows: $g \neq 0, T \neq 0$

When $g \neq 0$ and $T \neq 0$, the angles μ and β are again found as part of the solution. However, they can take, in principle, any values as it was the case in Sec. 6 where $T \neq 0$ and $g = 0$. This is to be contrasted to the case $g \neq 0$ and $T = 0$ of Sec. 7 where the angle μ was restricted to three values. We can therefore expect the limit $T \to 0$ to be a singular limit.

We present explicit results for the flow of Fig. 5. We first introduce the parameter

$$\alpha^* = \frac{\rho U^2 h}{T}. \tag{79}$$

Values of ν versus F for $\alpha^* = 10$ are presented in Figs. 14 and 15. These results show that there is a countably infinite set of values of F for which $\nu = 1$. This is to be contrasted to the case $T = 0$ (i.e., $\alpha^* = \infty$) for which $\nu = 1$ for all values $0 < F < F_c$. More interestingly it can be shown that this discrete set of values coalesce

J.-M. Vanden-Broeck

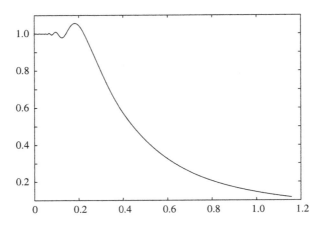

Fig. 14.   Values of ν versus F for $\alpha^* = 10$.

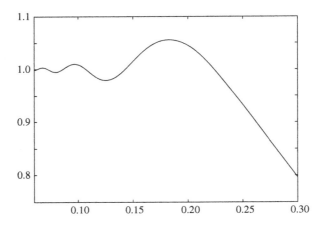

Fig. 15.   Enlargement of Fig. 14 showing clearly the oscillations.

to a unique value

$$F^{**} \approx 0.23 \quad \text{as } T \to 0, \tag{80}$$

(see Ref. 5). This is illustrated in Fig. 16 where we plot the value F_1^* of the largest value of the discrete set versus $1/\alpha^*$. As $T \to 0$ (i.e., as $1/\alpha^* \to 0$), $F_1 \to F^{**}$ in agreement with (80). The value of F^{**} is close to the experimental value (78). This shows that the physically relevant bubble is selected by including the surface tension T and taking the limit $T \to 0$.

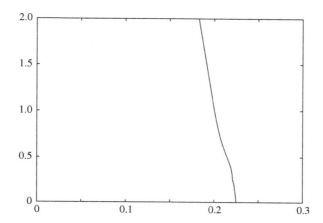

Fig. 16.   Values of $1/\alpha^*$ versus F_1^*. As $1/\alpha^* \to 0$, $F_1^* \to F^{**} \approx 0.23$.

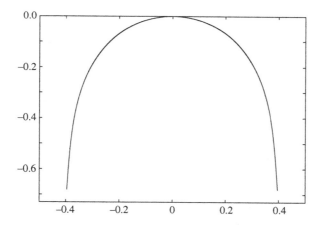

Fig. 17.   The selected bubble.

The selected profile is shown in Fig. 17.

9.   Some Exercises

(1) Derive equation (38). Hint: first map the strip of Fig. 6 onto to the lower half-plane.
(2) Derive equation (43).

(3) Use the result in (1) to derive (44).
(4) Define the contraction ratio y_F/y_E in Fig. 3. Consider now the flow in Fig. 8. Show that the contraction ratio is $\frac{\pi}{2+\pi}$.
(5) Derive the appropriate generalisation of (66) for the flow of Fig. 4. Assume $T = 0$ and $g \neq 0$.

10. Conclusions

We have studied some nonlinear gravity-capillary free surface flows. Special attention was devoted to problems for which the free surfaces intersect rigid surfaces. We hope to have convinced the reader of the mathematical beauty of these flows. Due to the space limitation, we restricted our attention to steady waveless potential flows. Further extensions to rotational flows, free surface flows with waves, time dependent problems and 3D flows can be found in Vanden-Broeck[5] and in the references cited there.

References

1. D. J. Acheson, *Elementary Fluid Dynamics*. Oxford University Press, UK, 1990.
2. G. K. Batchelor, *An Introduction to Fluid Dynamics*. Cambridge University Press, Cambridge, 2000.
3. G. Birkhoff and E. Zarantonello, *Jets, Wakes and Cavities*. Academic Press, New York, 1957.
4. M. Gurevich, *Theory of Jets and Ideal Fluids*. Academic Press, New York, 1965.
5. J.-M. Vanden-Broeck, *Gravity-Capillary Free Surface Flows*. Cambridge University Press, London, 2010.

Chapter 5

Internal Fluid Dynamics

Frank T. Smith

Department of Mathematics, UCL,
Gower Street, London WC1E 6BT, UK
f.smith@ucl.ac.uk

The chapter is on constriction/distortion and branching in a vessel or network of vessels containing fluid flow. The central interest is in medium-to-high Reynolds numbers where asymptotic approaches and matching yield insight in symmetric and non-symmetric cases over short or long length scales with or without viscous-inviscid interactions. Attention is given to side-branching, large networks, viscous wall layers, flow reversal, eddies, upstream influence and three-dimensional (3D) effects. A final discussion includes exercises and outlook.

1. Introduction

We consider internal fluid dynamics for constrictions, other distortions, branchings, networks involving channels and pipes. The practical interests are many, in industry, engineering, biomedical and environmental settings, where fluid-flow properties such as wall shear stress (WSS) and pressure (p^*) play vital roles. We seek mathematical progress and understanding, allying this with physical arguments, modelling, direct numerical simulation and experiments. The fluid is taken to be incompressible with constant density ρ^*, Newtonian, and its flow as laminar, steady or unsteady, two- or three-dimensional (2D, 3D). See Fig. 1. Clearly this is a vast area overall. We shall be mostly concerned with flows at medium-to-large Reynolds numbers Re for mathematical and physical interest and to provide checks and comparisons with simulations and experiments. Limitations in terms

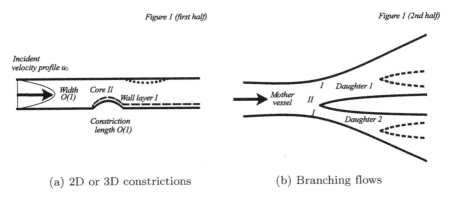

Fig. 1. Non-dimensional picture for (a) constriction (symmetric or not) and (b) branching from 1 to 2 or more vessels with typical $O(1)$ vessel widths.

of turbulence, 3D behaviour, flexible walls, non-Newtonian effects, wall roughness and asymptotic predictions may be touched on.

Section 2 below presents the equations of motion and the scales.[1–15] Section 3 describes theory and analysis for constrictions that are symmetric in 2D or axisymmetric in 3D. A gradual increase in the strength of constriction is discussed leading to a progression from strong to stronger to severe constriction in which flow reversal and separation can come into play significantly. Section 4 extends the arguments to motion inside branching vessels over comparatively short length scales and Sec. 5 to interactions between long and short length scales and/or between viscous and inviscid forces, again for branching flows, with attention on side-branching, large networks and viscous wall layers. Upstream influence is addressed in Sec. 6, especially with non-symmetry in 2D, while 3D effects for non-axisymmetric constrictions or branchings are the subject of Sec. 7. Final comments are made in Sec. 8, including suggested exercises and some model answers.

2. Equations and Scales

The flow velocity vector and components are $\boldsymbol{u}^* = (u^*, v^*, w^*)$, with corresponding Cartesian coordinates (x^*, y^*, z^*), the dynamic pressure p^*, the time t^*, as well as the uniform density ρ^*, viscosity μ^*

and kinematic viscosity $\nu^*(=\mu^*/\rho^*)$ of the fluid. An asterisk refers to a dimensional quantity. Concerning the governing equations, to rid ourselves of the effects of units we work in non-dimensional quantities, which provide more clarity. Thus we suppose a representative velocity u_r^* say, a representative length l_r^*, and then write

$$(x^*, y^*, z^*) = l_r^*(x, y, z), \boldsymbol{u^*} = u_r^* \boldsymbol{u}, p^* = \rho^* u_r^{*2} p, t^* = \left(\frac{l_r^*}{u_r^*}\right) t \quad (1)$$

with $\boldsymbol{u}$ denoting (u, v, w). Here u_r^* is the characteristic velocity amplitude of flow expected in the vessel and l_r^* the characteristic distance measured laterally across the vessel. The time factor is the convection-based time l_r^*/u_r^* and the pressure factor is the convection-based value $\rho^* u_r^{*2}$ for convenience. The governing equations then[15] are the non-dimensional continuity and Navier–Stokes equations in Cartesian coordinates,

$$\boldsymbol{\nabla} \cdot \boldsymbol{u} = 0, \quad (2)$$

$$\partial_t \boldsymbol{u} + (\boldsymbol{u} \cdot \boldsymbol{\nabla})\boldsymbol{u} = -\boldsymbol{\nabla}p + Re^{-1}\nabla^2 \boldsymbol{u} \quad (3)$$

in turn, where $\boldsymbol{\nabla} = (\partial_x, \partial_y, \partial_z)$ is the gradient operator. The boundary conditions are

$$\boldsymbol{u} = 0 \text{ on walls}, \quad (4)$$

$$\text{inflow (outflow) constraints far upstream (downstream).} \quad (5)$$

The requirement (4) is the no-slip condition holding on any fixed solid surface adjoining the fluid. Also, occasionally, we will use $u = \psi_y, v = -\psi_x$ instead of (2), if in 2D, where ψ is the streamfunction.

In (3), the Reynolds number $Re = u_r^* l_r^*/\nu^*$ measures the representative ratio of inertial forces to viscous forces. This is in line with the left-hand side of (3) representing inertia or convection along with acceleration, whereas viscous effects are on the right-hand side. In applications Re can range from large to small values, pointing to the use of reduced forms of the full system (2, 3).

The main reduced forms, for medium to large Re, are the Euler equations and the boundary layer (or wall-layer) equations. The former stem from simply neglecting the viscous contributions in (3); the scalings then are, for example,

$$\mid \mathbf{x} \mid \sim \mid t \mid \sim \mid \mathbf{u} \mid \sim \mid p \mid = O(1), \quad (6)$$

accompanied by a formal limit as $Re \to \infty$. An even better approach is to use asymptotic expansions: see later sections. Also the neglect of viscous contributions, the highest derivatives in (3), means we can only apply part of (4), the usual guess for this being the tangential-flow or zero-normal-velocity condition: if say the wall lies along $y = 0$ then $v = 0$ is required there, leaving u, w as usually non-zero. That guess is tantamount to assuming there is no substantial departure of the Euler flow from the wall, i.e., the motion is separation-free. This brings us to the viscous boundary layer equations which hold in relatively thin layers, typically between an Euler flow and a wall. The thinness accentuates the y-derivatives in the viscous terms and leads to the leading order pressure being independent of y. Again scaling and limiting apply. The boundary layer relies on a balance $|u|^2/|x| \sim Re^{-1}|u|/|y|^2$ being struck between inertial and viscous contributions, provided that the acceleration and pressure contributions respond passively to the others. If $|u|$ is typically $O(1)$ then we obtain the classical scaling[9]

$$|y| \sim Re^{-\frac{1}{2}}|x|^{\frac{1}{2}} \tag{7}$$

for the "thickness" of the relatively thin boundary layer.

Other significant subsets exist. One gives the lubrication layer, in which inertia terms are negligible compared with a boundary layer. By contrast, the main reduced form for small Re is Stokes equations, where the left-hand side of (3) is negligible. There are also Stokes layers and Rayleigh layers possible in which $u_t = Re^{-1}u_{yy}$ predominantly.

Concerning long and short scales, a full approach is expressed more systematically in terms of asymptotic expansions for specific contexts in Sec. 3 onwards. Major length scales we have seen are (i) $|x| \sim Re$, (ii) $|x| \sim 1$. The former (i) leads to the boundary layer equations acting across the whole vessel and, further downstream, the lubrication equations. These agree with the exact solution $u = u(y)$ in a straight or nearly straight channel which implies the continuity equation $0 = 0$ in effect and momentum equations

$$0 = -\frac{\partial p}{\partial x}(x) + Re^{-1}\frac{d^2 u}{dy^2}, \quad 0 = -\frac{\partial p}{\partial y} + 0. \tag{8}$$

The solution for u therefore gives the parabolic velocity profile

$$u(y) = \frac{1}{2}(y - y^2)\left(-Re\frac{dp}{dx}\right). \tag{9}$$

This exact solution represents plane Poiseuille flow (PPF) in a straight channel with $y = 0, 1$ at the walls, and it confirms that typically $|x|$ is $O(Re)$ if the imposed pressure difference is $O(1)$.

Concerning (ii), the $O(1)$ length scale which is inferred from the geometry and the vessel width is significant as it yields the Euler equations, hence ellipticity, hence upstream influence. "Jumps" can also be induced. Suppose locally in a 1-to-2 branching we have plug flow with the velocity profile $u_0(y)$ upstream in the mother vessel being uniform; then the vorticity curl $\boldsymbol{u}$ is zero throughout the local flow if separation-free and so constant velocities $u = u_1, u = u_2$ emerge in the two daughters downstream. See Fig. 1. With prescribed end-pressures $p = \pi_0, \pi_1, \pi_2$ in the three vessels, the values u_0, u_1, u_2 are unknowns. Integrals, however, yield the requirements of mass and momentum conservation,

$$u_1 h_1 + u_2 h_2 = u_0 h_0, \tag{10}$$

$$\pi_1 + \frac{1}{2}u_1^2 = \pi_2 + \frac{1}{2}u_2^2 = \pi_0 + \frac{1}{2}u_0^2 \tag{11}$$

which act as three algebraic equations controlling u_0, u_1, u_2. Solutions can be obtained readily or shown not to exist. More daughters, more generations or curved incident profiles are more subtle to treat and understand, a theme taken up later in the chapter. Also, one sees a picture of development from Euler to boundary layer to lubrication to PPF as downstream distance increases. Third, we have (iii), the numerous axial length scales between (i) and (ii). Here, a boundary or wall layer forms near any wall and an inviscid core in the rest of the flow field. The wall-layer thickness has scale

$$\text{either } |y| \sim Re^{-\frac{1}{2}}x^{\frac{1}{2}} \text{ or } |y| \sim Re^{-\frac{1}{3}}x^{\frac{1}{3}}, \tag{12}$$

depending on whether the incident velocity profile is $O(1)$ or $O(y)$ close to the wall, similarly to (7). Fourth (iv) are length scales lying outside the above ranges, giving lubrication (longer) or slow-flow (shorter) responses for instance.

3. Constrictions

We consider here the steady 2D motion through a symmetrically constricted channel when Re is asymptotically large, or equivalently through an axisymmetric pipe. For channel flows, we have conditions

$$u \to u_0(y), \quad v \to 0, \quad p \sim -\frac{2x}{Re} \quad \text{as } x \to -\infty, \qquad (13)$$

if PPF with $u_0(y) \equiv y - y^2$ applies far upstream between the undisturbed channel walls $y = 0, 1$; generalisations of these conditions are possible.

3.1. *Strong constriction*

A "strong constriction" is the first to produce a nonlinear response with unknown pressure, so that any separation, i.e., flow reversal, encountered is regular. If the typical constriction length is $O(1)$ the crucial thickness inside a symmetric channel is $O(Re^{-1/3})$ in y from (12). Thus, with the strong constriction given by $y = h\,Re^{-1/3}F(x)$ at the lower wall and $0 < h < \infty$, the flow near it, in the viscous wall layer I of Fig. 1, is described[1,14] by the asymptotic expansions

$$(u, \psi, p) = (Re^{-\frac{1}{3}}U, Re^{-\frac{2}{3}}\Psi, Re^{-\frac{2}{3}}P(x)) + \cdots, \quad y = Re^{-\frac{1}{3}}Y$$
$$x = O(1), \qquad (14)$$

in view of (2, 3) and the required inertia-pressure-viscous balance. Substituting into (2, 3) and keeping track of the scaling factors then gives

$$Re^{-\frac{1}{3}}U_x + \cdots - Re^{-\frac{1}{3}}\Psi_{xY} + \cdots = 0 \qquad (15)$$

in (2) since $v = -\psi_x$, while in (3) the x component has

$$Re^{-\frac{2}{3}}UU_x + \cdots - Re^{-\frac{2}{3}}\Psi_x U_Y + \cdots$$
$$= -Re^{-\frac{2}{3}}P_x + Re^{-1}Re^{\frac{1}{3}}U_{YY} + \cdots \qquad (16)$$

and the y component yields

$$O(Re^{-1}) = -Re^{-\frac{1}{3}}P_Y + \cdots. \qquad (17)$$

Hence, we are led to the boundary layer equations in normalised form

$$U = \Psi_Y, V = -\Psi_x, UU_x + VU_Y = -P'(x) + U_{YY}. \qquad (18)$$

The pressure $P(x)$ is unknown; thus, we have an interactive rather than a classical boundary layer. The boundary conditions are

$$(U, \Psi, P) \rightarrow \left(Y, \frac{1}{2}Y^2, 0\right) \quad \text{as } x \rightarrow -\infty \text{ [from (13)]}, \qquad (19)$$

$$U = \Psi = 0 \quad \text{at } Y = hF(x) \text{ [from (6)]}, \qquad (20)$$

$$U \sim Y + 0 \quad \text{as } Y \rightarrow \infty \left[\Rightarrow \Psi \sim \frac{1}{2}Y^2 + P(x)\right]. \qquad (21)$$

The condition on Ψ in (21) stems from integrating that on U, giving $\Psi \sim \frac{1}{2}Y^2 + Q(x)$ say and assuming $U - Y \ll Y^{-1}$ for this "condensed flow". So substitution into (18) leaves us with $Q' = P'$ as the non-zero terms at large Y. Hence, $Q = P + c_1$ where c_1 is a constant; but as $x \rightarrow -\infty$, $\Psi \rightarrow \frac{1}{2}Y^2$ corresponding to PPF upstream, and $P \rightarrow 0$ in (19); thence $Q(-\infty) = 0$, $c_1 = 0$, verifying the condition in (21).

The nonlinear problem (18)–(21) is a closed one, and solutions for $h \ll 1$ (weak constriction) and for $h = O(1)$ (strong constriction) are shown in Fig. 2, giving the pressure distribution $P(x)$; there is no upstream influence yet; also, $h = 0.6$ represents $h \ll 1$ well. For $h \ll 1$,

$$(U, \Psi, P) = \left(Y, \frac{1}{2}Y^2, 0\right) + h(\tilde{U}, \tilde{\Psi}, \tilde{P}) + \cdots \qquad (22)$$

and so (18) becomes, at order h, the linearised form

$$\tilde{U} = \tilde{\Psi}_Y, \tilde{V} = -\tilde{\Psi}_x, Y\tilde{U}_x + \tilde{V} = -\tilde{P}'(x) + \tilde{U}_{YY}. \qquad (23)$$

Taking a Fourier transform (FT) denoted by superscript f converts (23) to

$$\tilde{U}^f = \tilde{\Psi}^f_Y, \tilde{V}^f = -i\alpha\tilde{\Psi}^f, i\alpha Y\tilde{U}^f + \tilde{V}^f = -i\alpha\tilde{P}^f(\alpha) + \tilde{U}^f_{YY}, \qquad (24)$$

where α is the transform variable. Therefore, the shear stress perturbation $\tilde{\tau} \equiv \partial\tilde{U}/\partial Y$ satisfies, from differentiation in (23),

$$i\alpha Y\tilde{\tau}^f = \tilde{\tau}^f_{YY}. \qquad (25)$$

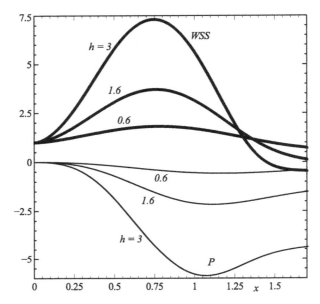

Fig. 2. Strong constriction: pressure P and WSS as strength h varies for shape $hx^2(2-x)^2$ in $0 \leq x \leq 2$, zero otherwise.

The appropriate solution bounded at infinity is then

$$\tilde{\tau}^f = \tilde{B}(\alpha)Ai(\xi) \text{ where } \xi \equiv (i\alpha)^{\frac{1}{3}}Y \tag{26}$$

with Ai denoting the Airy function. Using now a Prandtl transposition (PT) loosely we require $\tilde{\tau}^f_Y = i\alpha\tilde{P}^f$ at $Y = 0$, while (21) requires $\tilde{U}^f \to F^f$ as $Y \to \infty$, and so

$$(i\alpha)^{\frac{1}{3}}\tilde{B}Ai'(0) = (i\alpha)\tilde{P}^f, \quad (i\alpha)^{-\frac{1}{3}}\tilde{B}\kappa = F^f, \tag{27}$$

where $\kappa = \int_0^\infty Ai(q)dq = \frac{1}{3}$. It follows that

$$\tilde{P}^f = 3Ai'(0)(i\alpha)^{-\frac{1}{3}}F^f. \tag{28}$$

Convolution properties in the inverse FT therefore give the prediction

$$\tilde{P}(x) = -\gamma \int\limits_0^x F(s)(x-s)^{-\frac{2}{3}}ds \tag{29}$$

explicitly for the pressure produced by a given wall indentation $F(x)$ starting at $x = 0$. The constant $\gamma \equiv -3Ai'(0)/\Gamma(\frac{1}{3}) > 0$. The resulting pressure gradient is favourable as the fluid is pressed against the front of a constriction, but on the leeward side the pressure rises and the skin friction WSS falls rapidly. The trend continues as h increases. Nonlinearly, for sufficiently large h regular separation, i.e., flow reversal occurs[1, 14]: see Fig. 2 and exercises. After separation the pressure remains negative and its variation is quite small in the ensuing reversed flow, until reattachment takes place with the pressure P thereafter rising to its ultimate downstream value of zero.

Next, in the core II it is deduced that

$$(u, \psi, p) = (u_0(y), \psi_0(y), 0) + Re^{-\frac{2}{3}}(u_1, \psi_1, p_1) + \cdots \qquad (30)$$

for y of $O(1)$, where $\psi_0(y) \equiv \frac{1}{2}y^2 - \frac{1}{3}y^3$. A substitution process as in (15)–(17) shows that to leading order linearised Euler-flow properties hold in the form

$$u_0\psi_{1xy} - \psi_{1x}u_0' = -p_{1x} \quad and \quad -u_0\psi_{1xx} = -p_{1y}, \qquad (31)$$

so that on elimination of p_1 by cross-differentiation ψ_1 satisfies the linear elliptic equation

$$u_0\left(\frac{\partial^2\psi_1}{\partial x^2} + \frac{\partial^2\psi_1}{\partial y^2}\right) = u_0''\psi_1 \qquad (32)$$

with boundary conditions

$$\psi_1 = 0 \quad \text{at } y = \frac{1}{2} \text{ [for symmetry]}, \qquad (33)$$

$$\psi_1 \to 0 \quad \text{as } x \to -\infty \text{ [from (13)]}, \qquad (34)$$

$$\text{no exponential growth as } x \to \infty, \qquad (35)$$

$$\psi_1 \to P(x) \quad \text{as } y \to 0+ \text{ [to match with I]}. \qquad (36)$$

Here, (36) stems from (21) which gives

$$Re^{\frac{2}{3}}\psi \to \frac{1}{2}Re^{\frac{2}{3}}y^2 + P(x) \qquad (37)$$

with the y^2 term feeding into $\psi_0(y)$ and the $P(x)$ term feeding into (36). Solutions for ψ_1 are as in the next subsection.

We notice the small size of the core-flow disturbance in (30) and the absence of any nonlinear upstream response. For, if the constriction starts at $x = 0$ as we shall assume henceforth, then

$$U = Y, \qquad \Psi = \frac{1}{2}Y^2, \qquad P(x) = 0 \quad \text{for } x < 0. \tag{38}$$

The upstream response as yet is only linear and is due to the feedback from the ellipticity in the inviscid core-flow II. The solution of (32)–(36) gives the typical inviscid slip property

$$u_1 \to A_1(x) \quad \text{as } y \to 0+, \text{ in } x < 0 \tag{39}$$

in view of (36) with (38). The slip velocity or negative displacement $A_1(x)$ is a given function determined from (32)–(36). Hence, upstream $(x < 0)$ in the viscous layer in I again,

$$(u, \psi, p) = \left(Re^{-\frac{1}{3}}Y, \frac{1}{2}Re^{-\frac{2}{3}}Y^2, 0 \right)$$
$$+ (Re^{-\frac{2}{3}}U_1, Re^{-1}\Psi_1, Re^{-1}P_1(x)) + \cdots \tag{40}$$

and from substitution as in (15)–(17) again U_1, Ψ_1, P_1 satisfy linear viscous equations of motion with the outer boundary condition

$$U_1 \sim -Y^2 + A_1(x) \quad \text{as } Y \to \infty. \tag{41}$$

So the upstream wall-layer response is also linear and in particular upstream separation is not possible yet.

However, in the spirit of investigation, suppose we let h become large and look for the trends of the flow solutions in the regions I, II above. It can be shown[1] mainly from (21) that $P(x) \sim -\frac{1}{2}h^2(F(x))^2$ when h is large, at least if $F'(x) \geq 0$: see exercises. So then the core perturbation ψ_1 becomes larger like h^2, from the wall condition (36), implying the upstream slip velocity $A_1(x) = O(h^2)$ also. Therefore, in the upstream viscous response (40) the induced second-order perturbation $Re^{-2/3}U_1$ becomes $O(h^2Re^{-2/3})$, from (41), and so is expected to become comparable with the leading term $Re^{-1/3}Y$ of the velocity u when h grows to order $Re^{1/6}$, formally.

3.2. *Stronger constriction*

We consider next a symmetric "stronger" constriction of height $O(Re^{-1/6})$, given by $y = h_M Re^{-1/6} F(x)$, where now $0 < h_m < \infty$. The core flow II of Fig. 2 then suffers a larger perturbation than before, $Re^{-1/3}$ replacing $Re^{-2/3}$ in (30), but this leads again to (32)–(35) for ψ_1, although the wall constraint

$$\psi_1 \rightarrow \frac{1}{2} h_M^2 (F(x))^2 \quad \text{as } y \rightarrow 0+, \tag{42}$$

holds instead of (36). The constraint (42) follows either from the suggestion above that $P \sim -\frac{1}{2} h^2 F^2$ or simply as a Taylor series expansion of the typical inviscid condition $\psi \rightarrow 0$ as $y \rightarrow h_M Re^{-1/6} F(x) + \cdots$. The solution of (32)–(35) with (42) gives the upstream form (39) again, and so in the viscous wall layer I upstream

$$(u, \psi, p) = (Re^{-\frac{1}{3}} U, Re^{-\frac{2}{3}} \Psi, Re^{-\frac{2}{3}} P(x)) + \cdots \text{(for } x < 0 \text{ only).} \tag{43}$$

Hence, from (2, 3), U, Ψ, P satisfy the nonlinear problem of (18)–(21) except that

$$U = \Psi = 0 \quad \text{at } Y = 0, \tag{44}$$

$$U \sim Y + A_1(x) \quad \text{as } Y \rightarrow \infty, \tag{45}$$

since $F(x) = 0$ for $x < 0$. Here, we repeat, the negative displacement $A_1(x)$ is a known function of x.

Other details of the flow structure for this size of disturbance can be worked through but the upstream effect is the most vital for the following reasons. The displacement function $-A_1(x)$ is given by the infinite series[10]

$$-A_1(x) = \sum_{n=1}^{\infty} \kappa_n \exp(\gamma_n x), \tag{46}$$

where the constants κ_n depend on the specific constriction shape $h_M F(x)$ and the constants γ_n are the ordered eigenvalues of (32)–(34) with (42):

$$0 < \gamma_1 < \gamma_2 < \gamma_3 < \cdots, \qquad \text{e.g., } \gamma_1 = 5.175.$$

Solutions of (18) with (44)–(46) show that first, as $h_M \rightarrow 0$, a match with the structure for strong constriction is achieved; second, for

h_M of $O(1)$ and sufficiently large, upstream regular separation can occur, at $x = x_{\text{sep}}$ say; and, third, when $h_M \to \infty$, the core feedback through (45), (46) becomes so severe that this separation is pushed indefinitely far upstream, $x_{\text{sep}} \to -\infty$. In fact, the series in (46) gives us the asymptotic form

$$x_{\text{sep}} \sim -\frac{2}{\gamma_1} \ln(h_M) + O(1) \quad \text{as } h_M \to \infty, \tag{47}$$

since $\kappa_1 \propto h_M^2$. Thus, the separation far upstream is a form of free interaction, as the specific value of κ_1 affects only the $O(1)$ contribution or origin shift in (47). Far upstream for h_M large $-A_1(x) \sim \kappa_1 \exp(\gamma_1 x)$ and from the regular separation process then a pressure rise of order $Re^{-1/3}(P \text{ of } O(1))$ is produced overall, with a breakaway separation and structure like that of external flow[1] emerging downstream. So once again the upstream effect starts to change the whole flow structure.

3.3. *Severe constriction*

Here, the constriction is yet stronger, of height $O(1)$. Formally, for a "severe" constriction, h_M grows to $O(Re^{1/6})$ and so (47) implies separation at a large distance

$$-x = -x_{\text{sep}} = \frac{1}{3\gamma_1} \ln Re + O(1), \tag{48}$$

upstream.

The suggestion (48) is verified by a structural analysis[10] of severely constricted symmetric flows. Breakaway separation occurs both upstream, as the free interaction near (48), and on the constriction, via the incompressible triple-deck breakaway.[15] The Euler equations control the core of the motion, giving conservation of vorticity along streamlines there, but free streamline conditions of constant pressure hold along the unknown separated streamlines $\psi = 0$ present both upstream and downstream. The final downstream reattachment occurs far beyond the constriction, when x is $O(Re)$ and the boundary layer equations hold because y, u, p are all $O(1)$: see Sec. 2. Other relevant points include comparisons with experiments or Navier–Stokes calculations for axisymmetric pipe flow and symmetric channel flow. The comparisons are quite favourable.

4. Branchings

The 2D branchings in this section concern

$$|x| \sim 1 \tag{49}$$

and are governed mainly by the Euler system if separation-free, a matter discussed in Sec. 3 and below. There are many interesting features, for example, as summarised in Sec. 4.1, but we aim to focus more in Sec. 4.2 on features quite distinct from those in the previous section.

4.1. *Effects of vorticity, daughter numbers, nonlinearity*

These effects[1] include the following. Useful exact solutions are obtainable in certain branching cases such as with zero or uniform vorticity by conformal mapping. Next, the limit of many daughters yields an integral equation,

$$\int_0^1 \left\{ 2(\pi_0 - p(y)) + u_0^2 \right\}^{\frac{1}{2}} H(y) dy = u_0 h_0 \tag{50}$$

for the mother velocity $u_0(y)$ by extension from (10), (11). Finally here, the influence of one or more thin dividers of thickness $O(Re^{-1/3})$ in the core of a channel flow can be treated by analogy with Sec. S3 to give a small slip velocity u_w which acts as a scaled displacement effect added to the condition (21), with λ as incident WSS, namely

$$U \sim \lambda(Y + B(x)) \text{ as } Y \to \infty, \text{ with } B = \lambda^{-1} u_w(x) + hF(x) \tag{51}$$

which in turn acts to drive the linearised or nonlinear viscous-inviscid wall-layer response and can induce separation. We observe the latter shows a direct equivalence between branching and constriction for the small pressure changes involved. Two-phase flows with such dividers have been investigated recently, as have comparisons with direct numerical simulations.[6, 11]

4.2. *Substantial changes in cross-section*

This is for a branching geometry[12] similar to that in Fig. 1. A single mother tube, of width 1 and containing fully developed incident flow of unknown total mass flux $\propto \lambda$, branches locally into N daughter tubes of given total width $\bar{A}$ ("area") at large positive x; in the figure N is 4. The exit velocities u_1 to u_N in the daughters are unknown. The branching shape involves arbitrary $O(1)$ slopes and is prescribed, as are the daughter pressures π_1 to π_N downstream, which are measured relative to the upstream mother pressure taken as zero. The orders of magnitude point to an inviscid Euler response in the absence of significant separation.

Conservation of mass and pressure head apply in each daughter and so effectively ψ and $p + \frac{1}{2}(u^2 + v^2)$ are preserved. Hence, $\pi_n + \frac{1}{2}u_n^2 = 0 + \frac{1}{2}u_0^2$ for each n. Also, using an integration of $dy = d\psi/u$ with $u = u_n = (u_0^2(\psi) - 2\pi_n)^{1/2}$, incorporating (10) indirectly and summing over all the daughter tubes downstream yields the overall problem

$$\bar{A} = \sum_{i=1}^{N} \int_{\psi_i^-}^{\psi_i^+} \frac{d\psi}{\{u_0^2(\psi) - 2\pi_i\}^{\frac{1}{2}}}, \tag{52}$$

to determine the size λ of the incident stream function profile $\psi_0(= \lambda(\frac{1}{2}y^2 - \frac{1}{3}y^3))$ together with the stream function values $\psi_i^{\pm}$ on the walls of each daughter. The squared term in (52) can also be written $\psi_0'(\psi_0^{-1}(\psi))^2$. Properties of the solutions in Fig. 3 include: the trend $\bar{A} \to 1-$, which corresponds to the π_i becoming negligible in (52) since the typical velocities are large; the asymptotes at low $\bar{A}$ where $\lambda \to 0$ and the u_0^2 term in (52) becomes negligible; the cut-off to the right in some cases; and the linear growths in λ with increasing daughter numbers, due again to π_i becoming negligible. See exercises.

As part of a larger network, the local pressure differences π_i interact with longer-scale behaviour. Further, the 3D version remains a mystery: see Sec. 7. Meanwhile, comparisons with direct simulations again seem supportive.

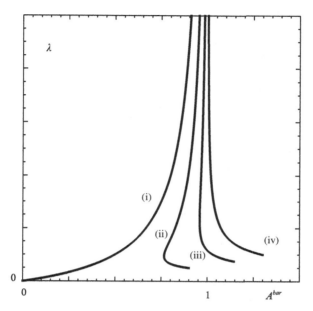

Fig. 3. Induced mass flux $\propto \lambda$ versus flow area $\bar{A}$ downstream, as N varies for different pressure drops (i)–(iv) in 2D multi-branching

5. Long-Short Interactions

The 2D branchings here have feedback between long and short scales.

5.1. *Side-branching*

Here, a relatively small daughter vessel D side-branches at location $x = x_1$ from the mother M. So on the scale of D only the incident uniform-shear profile matters at leading order. Direct simulations[7] for this are up to Re values of about 200. We naturally ask whether theory for large Re might help understanding. Over the length scale

$$|x| \sim 1 \tag{53}$$

the viscous-inviscid wall layer has thickness $|y| \sim Re^{-1/3}$ as in (12) and the governing equations are (18) subject to

$$U = V = 0 \text{ on all solid boundaries of } M,$$
$$D \text{ and } P \to P_\infty \text{ at } x = x_\infty \text{ in } D \tag{54}$$

and the conditions (19)–(21) in M. Here, P_∞ is the prescribed pressure downstream in D while that for upstream in M is zero, with the wall-layer system applying in M close to its wall and throughout D.

A significant issue concerns how much leakage there is into D. This creates something of a puzzle perhaps. The system (18) is parabolic in x yet we wish to impose the downstream requirements in (54) as well as the upstream one (19). The length-scale arguments of the previous sections provide the resolution, however, we can expect an Euler region to be relevant over the shorter scale

$$|x - x_1| \sim Re^{-\frac{1}{3}}, \tag{55}$$

to balance axial and lateral effects. The nonlinear adjustment gives a jump in velocity and pressure from M at $x = x_1-$ to D at $x = x_1+$ as seen from the scale of (53), (55). During the jump generally the pressure falls because of the zero wall velocity upstream. The jump structure only works, i.e., is non-trivial, at the entrance to D and it conserves quantities as in (10), (11), (50), (52); it provides the necessary upstream influence.

Hence, the calculation required involves the following:

(i) March forward in M to x_1, yielding $p(x_1-)$;
(ii) Guess the mass flux (the leakage) into D;
(iii) Derive $p(x_1+)$ from the jump and flux conditions;
(iv) March in D to x_∞;
(v) Test there on $p - p_\infty$;
(vi) Update the guess of mass flux, return to (iii) and iterate repeatedly until convergence.

Results[7] agree with direct simulations. Also, if $x_\infty \gg 1$, then most of the flow in D is as if far downstream and so yields lubrication behaviour (Sec. 2), in which case it is found that many of the results in the calculation can be replaced by formulae.

5.2. Larger networks

A large network[1] may comprise successive generations of bifurcations. The two major length scales are $O(1)$ and $O(Re)$, and the pressures are imposed only at the downstream and upstream ends of the network, not in between, and so the pressures π_n in Sec. 4.2. effectively are unknown in advance. The network leads to interesting nonlinear interactions between viscous thin-layer equations and inviscid-jump equations.

5.3. *Wall layers*

Viscous-inviscid interaction as before occurs in many of the wall layers present, notably those of Secs. 4, 5.1 above and 6 below. The general features of wall layers are exemplified by those cases, in all of which the boundary layer or wall-layer equations (18) play an important part. The pressure remains unknown in advance or adjustable in each case.

6. Non-symmetry in 2D

Non-symmetric effects can be significant for constrictions or branchings.

6.1. *Long or short non-symmetric constrictions*

Given a non-symmetric constriction of length scale $L \gg 1$ in a channel we can anticipate regions I–III because of Sec. 3. The y-scale in the thin wall layers I and III is $O(Re^{-1/3}L^{1/3})$, the u-scale is the same by virtue of the incident velocity profile near each wall, and the p-scale is therefore $O(Re^{-1/3}L^{1/3})$. Necessarily, $L \ll Re$. So in the core II where $y \sim 1$ the expansions are

$$u = u_0(y) + Re^{-\frac{1}{3}}L^{\frac{1}{3}}u_1(x,y) + \cdots, \tag{56}$$

$$v = Re^{-\frac{1}{3}}L^{-\frac{2}{3}}v_1(x,y) + \cdots \text{[from continuity]}, \tag{57}$$

$$p = Re^{-\frac{2}{3}}L^{\frac{2}{3}}p_1(x,y) + \cdots \text{ [from I, III]} \tag{58}$$

with $x = LX$. Substitution into (2), (3) gives at leading order

$$u_{1X} + v_{1y} = 0, \quad u_0 u_{1X} + v_1 u_0' = 0, \tag{59}$$

from continuity and axial momentum, whereas the normal momentum balance becomes

$$u_0 v_{1X} Re^{-\frac{1}{3}} L^{-\frac{5}{3}} + \cdots = -p_{1y} Re^{-\frac{2}{3}} L^{\frac{2}{3}} + \cdots, \tag{60}$$

predominantly. Hence, $p_{1y} = 0$ holds true in the core provided $Re^{-\frac{1}{3}} L^{-5/3} \ll Re^{-2/3} L^{2/3}$, that is

$$Re^{\frac{1}{7}} \ll L \ [\ll Re], \tag{61}$$

representing a range of long length scales. [The bracketed restriction stems from just before (56).] In such cases, the core yields

$$u_1 = A(X)u_0'(y), v_1 = -A'(X)u_0(y), p_1 = P(X). \tag{62}$$

Meanwhile, the lower wall layer I has y of order $Re^{-1/3} L^{1/3}$, giving

$$u = Re^{-\frac{1}{3}} L^{\frac{1}{3}} U + \cdots, \tag{63}$$

$$v = Re^{-\frac{2}{3}} L^{-\frac{1}{3}} V + \cdots, \tag{64}$$

$$p = Re^{-\frac{2}{3}} L^{\frac{2}{3}} P + \cdots \tag{65}$$

and this substituted into (2), (3) implies we have to solve (18) subject to

$$U = V = 0 \quad \text{at } Y = hF(X), \tag{66}$$

$$U \sim Y + A(X) \quad \text{as } Y \to \infty. \tag{67}$$

Here, F is the scaled lower wall shape. Similarly the upper wall layer III has $y - 1$ of order $Re^{-1/3} L^{1/3}$, with the asymptotic expansions

$$u = Re^{-\frac{1}{3}} L^{\frac{1}{3}} \bar{U} + \cdots, \tag{68}$$

$$v = Re^{-\frac{2}{3}} L^{-\frac{1}{3}} \bar{V} + \cdots, \tag{69}$$

$$p = Re^{-\frac{2}{3}} L^{\frac{2}{3}} \bar{P} + \cdots \tag{70}$$

which yield (18) but for $\bar{U}, \bar{V}, P$, since p_1 is independent of y and $\bar{P}(= P)$ is independent of $Y, \bar{Y}$. The boundary conditions are

$$\bar{U} = \bar{V} = 0 \text{ at } \bar{Y} = hG(X), \tag{71}$$

$$\bar{U} \sim -\bar{Y} - A(X) \text{ as } \bar{Y} \to -\infty, \tag{72}$$

as well as $(\bar{U}, \bar{V}, P) \to (-Y, 0, 0)$ as $X \to -\infty$; G is the upper wall shape.

Applying PTs, $Y - hF(X) \equiv Y_1, V - hF'(X)U \equiv V_1$ in I and $\bar{Y} - hG(X) = -\bar{Y}_1, \bar{V}_1 - hG'(X)\bar{U} \equiv -\bar{V}_1$ in III transforms the conditions (66, 67) to

$$U = V_1 = 0 \quad \text{at } Y_1 = 0, \quad U \sim Y_1 + (A + hF) \quad \text{as } Y_1 \to \infty \quad (73)$$

and the conditions (71, 72) to

$$\bar{U} = \bar{V}_1 = 0 \quad \text{at } \bar{Y}_1 = 0, \quad \bar{U} \sim \bar{Y}_1 - (A + hG) \quad \text{as } \bar{Y}_1 \to \infty, \quad (74)$$

while leaving the governing equations unchanged except for replacement of $(Y, V, \bar{Y}, \bar{V})$ by $(Y_1, V_1, \bar{Y}_1, \bar{V}_1)$. Comparison of (73), (74) then points to a solution in which $(U, V_1), (\bar{U}_1, \bar{V}_1)$ are identical and $(A + hF) = -(A + hG)$, implying[1,5,8,14]

$$A = \frac{1}{2}h(F + G). \quad (75)$$

This simple result that the displacement in the core is the average of the upper and lower wall displacements leads to the flow problems in layers I, III being identical. Indeed, they repeat the problem we had in Sec. 3 for symmetric constriction, with now in effect

$$U = V = 0 \quad \text{at } Y = 0, \quad U \sim Y + \frac{1}{2}(F - G) \quad \text{as } Y \to \infty. \quad (76)$$

Comments are the following. (a) Streamlines in the core take the "average path" between the given wall constriction shapes. (b) If $F = G$ then the whole flow rides up and over as a displaced PPF in which $(U, V, P) \equiv (Y, 0, 0)$ throughout. (c) If $F = -G$ then we retrieve the symmetric case. (d) Examples of solutions[1,14] include some with eddies of reversed flow. (e) Concerning (61), if the length scale L approaches $O(Re)$ then layers I–III merge and so viscous effects matter across the whole cross-section, whereas if L decreases towards $O(Re)^{1/7}$ then the normal pressure gradient $\partial p/\partial y$ within the core becomes increasingly significant as in (60), a feature explored further in the next subsection; even shorter lengths are studied there too. (f) All cases before this have been for symmetric flows effectively.

6.2. *Upstream influence and instability*

What happens when

$$L \sim Re^{\frac{1}{7}}, \tag{77}$$

is that we find upstream influence. Putting $L = Re^{1/7}$ in (60) shows the core having

$$(u, v) = \left(u_0(y) + Re^{-\frac{2}{7}} A(X) u_0'(y), -Re^{-\frac{3}{7}} A'(X) u_0(y) \right) + \cdots, \tag{78}$$

$$p = Re^{-\frac{4}{7}} p_1(X, y) + \cdots \tag{79}$$

with $u_0 v_{1x} = -p_{1y}$ from y-momentum. Hence,

$$p_1(X, y) = A''(X) \int_0^y u_0^2(\hat{y}) d\hat{y} + P(X), \tag{80}$$

since $v_1 \equiv -A'(x) u_0(y)$. Therefore, our task is to solve

$$\text{[lower layer] } U_T + U U_X + V U_Y = -P_X(X, T) + U_{YY}, \tag{81}$$
$$\text{[upper] } \bar{U}_T + \bar{U} \bar{U}_X + \bar{V} \bar{U}_{\bar{Y}} = -\bar{P}_X(X, T) + U_{\bar{Y}\bar{Y}}, \tag{82}$$

along with: continuity equations; no-slip conditions at $Y = 0, \bar{Y} = 0$ after a PT; matching conditions $U \sim Y + (A + hF)$ as $Y \to \infty$ and $\bar{U} \sim \bar{Y} - (A + hG)$ as $\bar{Y} \to \infty$; and

$$\bar{P}(X, T) - P(X, T) = I A_{XX}(X, T). \tag{83}$$

The positive constant $I \equiv \int_0^1 u_0^2 dy$ is $\frac{1}{30}$ if u_0 is $y - y^2$. We have taken the opportunity here to add in unsteadiness as a slow $\partial_t \sim Re^{-3/7} \partial_T$ which enters the two wall layers explicitly but not the core.

For $h \ll 1$, a linearised analysis holds. With (81), the analysis proceeds as in (22), (23), so that formally

$$\tilde{\tau}_T + Y \tilde{\tau}_X = \tilde{\tau}_{YY}. \tag{84}$$

The FT converts ∂_X to $i\alpha$. Also we let ∂_T be replaced by $-i\omega$ to allow for $\exp(-i\omega T)$ temporal dependence of scaled frequency ω. Hence $i(-\omega + Y\alpha)\tilde{\tau}^f = \tilde{\tau}_{YY}^f$, schematically. Setting $\xi \equiv (i\alpha)^{1/3}(Y - Y_0)$ with

$Y_0 \equiv \omega/\alpha$ then yields Airy's equation $\xi \tilde{\tau}^f = \tilde{\tau}_{\xi\xi}^f$ and so analogously to (26) we have

$$\tilde{P}^f = \kappa^{-1} Ai'(\xi_0)(i\alpha)^{-\frac{1}{3}}(\tilde{A}^f + F^f), \tag{85}$$

which relates $\tilde{P}$ to $\tilde{A}, F$. Here, $\xi_0 \equiv -(i\alpha)^{1/3}\omega/\alpha, \kappa \equiv \int_{\xi_0}^{\infty} Ai(q)dq$ are complex constants. With (82), similar working leads to

$$\tilde{\tilde{P}}^f = \kappa^{-1} Ai'(\xi_0)(i\alpha)^{-\frac{1}{3}}(-\tilde{A}^f - G^f). \tag{86}$$

Finally here, the core effect (83) gives

$$\tilde{\tilde{P}}^f - \tilde{P}^f = -I\alpha^2 \tilde{A}^f, \tag{87}$$

leaving us with three equations (85–87) for $\tilde{P}^f, \tilde{\tilde{P}}^f, \tilde{A}^f$.

Subtracting (85, 86) and using (87) yields the negative displacement solution

$$\tilde{A}^f = \frac{-(F+G)^f}{\left[\frac{\kappa}{Ai'(\xi_0)}(i\alpha)^{\frac{7}{3}}I + 2\right]}, \tag{88}$$

in particular. The main points about (88) are as follows.

- If the length scale $\gg Re^{1/7}$ then the term "+2" dominates in the denominator and so we retrieve (75).
- Conversely for lengths $\ll Re^{1/7}$ in essence α is large, making $\tilde{A}^f$ relatively small. Thus, $\tilde{A}^f$ becomes negligible in (85) and the earlier result (28) is obtained for the lower-wall pressure. (In steady flow $\kappa = \frac{1}{3}$ and $\xi_0 = 0$.) Likewise $\tilde{A}^f$ becomes minor in (86), leaving (28) again but for the upper-wall pressure in terms of the upper-wall shape G^f. The two wall layers act independently, as if with zero displacement ("condensed flow"), while the core result (87) determines the small displacement correction. This continues for $Re^{-1/2} \ll L \ll Re^{1/7}$.
- In steady flow, the denominator identifies a pole in the lower half of the complex-α plane at

$$\alpha = -i\sigma \text{ where } \frac{\sigma^{\frac{7}{3}}I}{(-3Ai'(0))} = 2. \tag{89}$$

Recall $Ai'(0) < 0$. This is significant because of contours used in the FT inversion. For $X > 0$, a keyhole contour in $Im(\alpha) > 0$ is

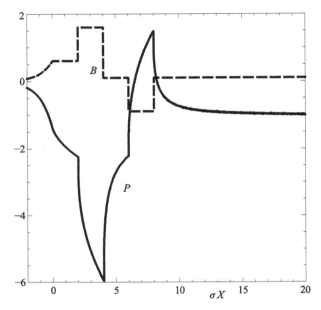

Fig. 4.   Network with upstream influence: pressures P and effective thicknesses B through successive bifurcations at $x = 0, 1, 2, 3, 4$.

relevant including a branch cut for $(i\alpha)^{1/3}$ since the constriction is taken to start at $X = 0$. For $X < 0$, a large semicircular contour is relevant in $Im(\alpha) < 0$. The Cauchy Residue Theorem therefore gives only a contribution from the pole (89) for $X < 0$, producing $\exp(i\alpha X)$ and hence

$$A, P, \bar{P} \propto e^{\sigma X}, \text{ in } X < 0, \tag{90}$$

apart from multiplicative constants. A solution is presented in Fig. 4, confirming the *upstream influence* now present.[1,5,8]

• In unsteady flow, the denominator in (88) identifies a resonance or *instability*. In fact,

$$\frac{\int_{\xi_0}^{\infty} Ai(q)dq}{Ai'(\xi_0)} = \frac{-2}{(i\alpha)^{\frac{7}{3}}I} \tag{91}$$

is the dispersion relation for instabilities in channel flow, corresponding to Tollmien-Schlichting ones in boundary layers.[15] There is a neutral value from (91) such that ω, α are both real, and temporal growth occurs for wavenumbers α above the neutral value.

6.3. *Branching networks*

Upstream influence over the length scale (77) similarly arises ahead of a non-symmetric bifurcation,[1] corner (where strong interaction leads to separations[5]) or branching network.[1] See Fig. 4.

Suppose first that we have a single 1-to-2 branching. The inviscid core then within the lower daughter D_1 acts mostly as if distinct from that in the upper daughter D_2 and likewise for the viscous upper wall layer, over the present length scale (77). In the D_1 core, the pressure is of order $Re^{-4/7}$ and the streamfunction expands as

$$\psi = \psi_0(y) + Re^{-\frac{2}{7}}\{A(X)u_0(y) + \lambda_2\psi_0(y)\} + \cdots, \qquad (92)$$

where the constant λ_2 is an unknown associated with the altered mass flux. The undeveloped viscous layers on the internal divider of D_1, D_2 have negligible impact (they are passive as in Sec. 2), implying a tangential-flow condition on the given divider underside $y = c_0 - Re^{-2/7}T_0(X)$ say. Taking $T_0(0)$ as zero without loss of generality thus yields the classical thin-channel result

$$A(X) = T_0(X) + K_0, \quad \text{for } X > 0, \qquad (93)$$

(since $u_0(c_0)$ is non-zero) which determines the function $A(X)$ to within the additive constant K_0. Similarly, upstream influence present in the mother tube M yields a free-interaction behaviour with unknown constant K,

$$A(X) = Ke^{\sigma X}, \quad \text{for } X < 0, \qquad (94)$$

as in (90), representing an elliptic effect. (This problem of the 1-to-2 case has also been treated by a Wiener–Hopf technique.) A feature due to the presence of the bifurcation however is that an axial jump in displacement can occur across the D_1, D_2 entrances from 0− to 0+, as in earlier examples. The jump is admissible and in fact necessary due to the set pressures upstream and downstream. At the outer walls in particular, where the incident velocity is close to zero, the viscous layers allow the Bernoulli quantity $p + u^2/2$ to be conserved along each local inviscid streamline as in Sec. 5.1 by means of a scaled pressure jump, in this case $\lambda_0^2(K^2 - K_0^2)/2$. The jumps are smoothed out over a shorter axial scale by an Euler region of length $O(1)$ in x which provides direct communication between D_1, D_2, M. The feature that K, K_0 are unequal in general allows adjustment of K_0 in order to allow the D_1 pressure to satisfy the downstream pressure condition, and likewise for D_2.

Second, suppose a 1-to-4 network. Then another new feature appears. Again attention can be restricted to a lower part, consisting now of a D_1 described earlier on and two granddaughters G_1, G_2 which begin at $X = X_1 > 0$. The lower G_1 is also described essentially as before but G_2 must suffer higher typical pressure variations of order $Re^{-2/7}$ such that

$$\Psi = \Psi_0(y) + Re^{-\frac{2}{7}}\{D(X)u_0(y) + \lambda_2[\Psi_0(y) - \Psi_0(c_1)]\} + \ldots, \quad (95)$$

where $c_1 - Re^{-2/7}T_1(X)$, $c_1 + Re^{-2/7}S_1(X)$ are the underside and topside respectively of the divider between G_1, G_2 and

$$D(X) = -p_1(X)\int_{c_1}^{y} u_0^{-2}dy - S_1(X) + \gamma_1. \quad (96)$$

The $Re^{-2/7}$ scaled pressure p_1 and the constant γ_1 are unknown. The novel feature here is that another jump must usually occur, namely, in pressure across the entrance of G_2 from X_1- to X_1+. This again is admissible, as the incident velocity is non-zero at all y heights of G_2, allowing the Bernoulli property to be maintained along each streamline. This active jump is also smoothed out on a shorter axial scale by an $O(1)$ Euler region in $X - X_1$. (Overall this is another type of ellipticity.) As a result, it is found that a jump is also induced in the effective $A(X)$ function here which although still similar to (93, 94) now has

$$A(X) = Ke^{\sigma X}, \text{ (discontinuity)}, \quad T_0(X) + K_0, \text{ (discontinuity)},$$
$$T_1(X) + K_1. \quad (97)$$

This doubly discontinuous form then drives the viscous wall-layer response by means of a constraint similar to (67). The displacement constants K_0, K_1 in (97) are controlled not only by the outermost

(G_1) imposed pressure downstream but also by the inner (G_2) grand-daughter pressure imposed downstream.

Larger networks produce similar effects,[1] i.e., potentially many discontinuities in the negative displacement $A(X)$ which, along with shape $f(X)$, forces the viscous layer by means of (67) and induces discontinuities in the wall pressure(s). The viscous layer is nonlinear in general, requiring numerical solution and admitting separation as in Sec. 3. By virtue of PT, the solution depends only on the effective thickness $(A+f)[= B$ say], thus giving wide application. For small B, a linearised form gives small discontinuities in pressure as in Fig. 4. A contraction of the outermost tube width broadly leads to a favourable pressure gradient and increasing wall shear, and expansion to an adverse pressure gradient with decreasing wall shear, as expected, but the discontinuities due to the branching junctions can counteract those trends.

7.   Non-symmetry in 3D

Internal flows in 3D are usually quite difficult to handle theoretically and the corresponding research area remains at the frontier of understanding. For constriction, there has been fair progress. For branchings, part of the reason for the difficulty in theoretical progress is simply that in 3D fluid particles can move about so much more than in 2D. Thus, within Sec. 4, for example, we cannot readily say in advance whether a given fluid particle entering in the incident mother-vessel flow will end up inside daughter one or daughter two far downstream. Nor can we say how much rotation there will be in the cross-sectional part of the motion in either of the daughters downstream. So analogues of the relations (10), (11), (50), (52) are not obtained readily.

7.1.   *3D Constrictions*

Distortions involving θ-dependence are most interesting here since otherwise axisymmetic flow is obtained which is essentially the same as for 2D channels. Non-symmetric confined bumps, dents, corners, curvatures and injections in 3D may all benefit in terms of under-standing predictions of pressure and WSS.

We begin with steady constricted flow.[1] The first crucial size of 3D constriction, in an otherwise straight pipe of circular cross-section ($r = 1$), is of height $O(Re^{-1/3})$ in the radial (r) direction with the length scale $|x|$ being $O(1)$: thus, we anticipate interesting effects when

$$|x| \sim |\theta| \sim 1. \tag{98}$$

Then the 3D equations governing the viscous wall-layer response are

$$U_x + V_Y + W_Z = 0, \tag{99}$$

$$UU_x + VU_Y + WU_Z = -P_x + U_{YY}, \tag{100}$$

$$UW_x + VW_Y + WW_Z = -P_Z + W_{YY} \tag{101}$$

with the streamwise, radial and azimuthal velocities expanded as

$$[u, v, w] = [Re^{-\frac{1}{3}}U, -Re^{-\frac{2}{3}}V, Re^{-\frac{1}{3}}W] + \cdots, \tag{102}$$

while Y, Z stand for $Re^{1/3}(1 - r), \theta$ in turn and the pressure is $Re^{-2/3}P(x, \theta) + \dots$. The no-slip condition applies to U, V, W at the scaled constriction surface $Y = hf(x, \theta)$ along with the outer condition

$$U \sim Y + 0, \quad W \to 0 \quad \text{as } Y \to \infty \tag{103}$$

and $(U, V, W, P) \to (Y, 0, 0, 0)$ as $x \to -\infty$. The lack of displacement for U in (103) is necessary for consistency in the inviscid core flow where the oncoming motion, e.g., the Poiseuille form $u = (1 - r^2)/2$, is only slightly disturbed as in Sec. 3.1. The constriction surface is given by $r = 1 - Re^{-1/3}hf(x, \theta)$, the PT changes the boundary conditions to

$$U = V = W = 0 \text{ at } Y = 0 \text{ and } U \sim Y + hf, W \to 0 \text{ as } Y \to \infty \tag{104}$$

in effect, and the flow solution must have period 2π in θ. Unsteady behaviour has $\partial U/\partial T, \partial W/\partial T$ added to (100), (101) respectively.

Linearised solutions[1] hold for small h, where

$$U = Y + h\tilde{U} + \cdots, (V, W, P) = h(\tilde{V}, \tilde{W}, \tilde{P}) + \cdots. \tag{105}$$

The system therefore reduces at $O(h)$ to

$$\tilde{U}_x + \tilde{V}_Y + \tilde{W}_Z = 0, \tag{106}$$

$$Y\tilde{U}_x + \tilde{V} = -\tilde{P}_x + \tilde{U}_{YY}, \tag{107}$$

$$Y\tilde{W}_x = -\tilde{P}_Z + \tilde{W}_{YY}, \tag{108}$$

$$\tilde{U} = \tilde{V} = \tilde{W} = 0 \quad \text{at } Y = 0 \text{ and } \tilde{U} \to f, \quad \tilde{W} \to 0 \quad \text{as } Y \to \infty \tag{109}$$

with $\tilde{P}(x, Z)$ unknown. Applying the FT in $(x, Z) \to (\alpha, \beta)$ or taking $\cos \beta Z, \sin \beta Z$ dependence for periodicity then produces the pressure and WSS solutions (see exercises) which exhibit a pole at $\alpha = -i\beta$ and thus *upstream influence*.

The presence of new upstream influence can also be seen from working in real space. Take the x-derivative of (100) and the Z-derivative of (101), add together, and we obtain

$$Y(\tilde{U}_{xx} + \tilde{W}_{xZ}) + \tilde{V}_x = -\nabla^2 \tilde{P} + (\tilde{U}_x + \tilde{W}_Z)_{YY} \tag{110}$$

with $\nabla^2 \equiv \partial_x^2 + \partial_Z^2$. So setting $\tilde{U}_x + \tilde{W}_Z = u^+, \tilde{V}_x = v^+$ now implies

$$u_x^+ + v_Y^+ = 0, \quad Yu_x^+ + v^+ = -p_x^+ + u_{YY}^+, \tag{111}$$

where the first part comes from (99) and the second from (110), while

$$\nabla^2 \tilde{P} = p_x^+. \tag{112}$$

The form (111) and its associated boundary conditions are quasi-2D, exactly as (18)–(21), and so can be solved for u^+, v^+, p^+ without upstream influence appearing; but then, afterwards, (112) acts to determine the real pressure $\tilde{P}$ by means of which upstream influence is provoked. (The Laplacian corresponds to a factor $(\alpha^2 + \beta^2)$ in exercise $E2$ below.)

Nonlinear cases[1, 3, 4, 13] can also be tackled by use of the so-called skewed shears u^+, v^+ combined with quasi-2D and 3D multi-sweeping. Moreover, analogous ideas apply in external flows. The feature of nonlinear upstream influence makes the investigation of stronger 3D constrictions interesting.

7.2.   *3D branchings over long or short scales*

For long scales, see elsewhere.[1] For short scales, our concern is with

$$|x| \sim 1 \tag{113}$$

in 3D. Such 3D non-symmetry is obviously the most common case in reality and in experiments and can be caused by unequal pressures at the downstream ends of the daughter vessels, by the branching geometry itself or by the incident velocity profiles in the mother being non-symmetric.

Arguments as in Sec. 7.1 suggest we concentrate first on flow rates for which

$$1 \ll Re \ll |p|^{-\frac{3}{2}}. \tag{114}$$

In brief, the configuration for pressure-driven flows applies for two or more daughters where the flow is driven merely by pressure differences, so that shape-effect forcings from the core and the outer wall are absent. Steady linearised flow is assumed for now. The wall-layer problem inferred from Sec. 7.1 has an unusual solution then in that $\tilde{V}$ is identically zero throughout, yielding Laplace's equation

$$\nabla_2^2 P = 0 \tag{115}$$

for $P(x, \theta)$, where ∇_2^2 denotes the 2D Laplacian $\partial_x^2 + \partial_\theta^2$ from (113). The boundary conditions prescribe P far upstream and downstream. The reduction of the fully 3D wall-layer problem to the 2D Laplacian problem is notable.

In a basic case of two daughters, $K = 2$, with end pressures P_1, P_2 such that $P_2 = -P_1$ there is an exact solution from a conformal mapping, such that

$$\frac{P}{P_1} = \pm \text{ real } (1 + e^{-2\varsigma})^{-\frac{1}{2}}, \tag{116}$$

where ς denotes $x + i\theta$ and the branching at $x = 0$ has the junctions of the outer walls and the divider being symmetrically disposed at $\theta = \pi/2, 3\pi/2$. The solution exhibits the irregular response of P near the leading edge of the divider in 3D, a response which is more severe than in the symmetric-flow setting despite the geometric shapes studied being identical. There is also the series representation

$$\frac{P}{P_1} = e^x \cos\ \theta - \left(\frac{1}{2}\right) e^{3x} \cos\ 3\theta$$

$$+ \left(\frac{3}{8}\right) e^{5x} \cos\ 5\theta + \cdots \quad \text{for } x < 0, \tag{117}$$

$$\frac{P}{P_1} = 1 - \left(\frac{1}{2}\right) e^{-2x} \cos\ 2\theta$$

$$+ \left(\frac{3}{8}\right) e^{-4x} \cos\ 4\theta + \cdots \quad \text{for } x > 0 \tag{118}$$

which shows the P eigenforms clearly. Here, (117) and (118) apply in the daughter with P_1, whereas the pressure in the other daughter is equal and opposite. The wall-pressure responses are presented in Fig. 5.

For more than two daughters, $K > 2$, conformal mapping or series solutions again apply. The wall pressure solutions P may be deduced for a range of values of K, along with velocity profiles and the induced

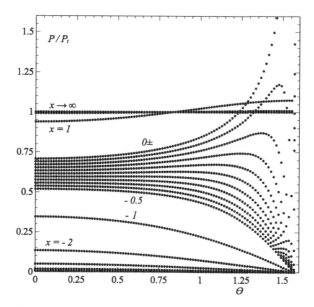

Fig. 5.   Branching in 3D: P versus θ for varying x with prescribed pressures 0 upstream, P_1 downstream; one mother, 2 daughters; divider starting at $x = 0$ is given by $\theta = \pm\pi/2$.

WSS distributions from (111), with λ constant, T dependence being absent and with P found as just above. For instance, the integral form

$$\tau_1 = \phi \int_{-\infty}^{x} (x - \xi)^{-\frac{2}{3}} \frac{\partial P}{\partial \xi}(\xi, \theta) d\xi, \quad \text{with } 2\pi\lambda^{\frac{1}{3}}\phi = -3^{\frac{1}{6}}\left(\Gamma\left(\frac{2}{3}\right)\right)^2,$$

(119)

$(\lambda^{\frac{1}{3}}\phi = -0.35047\ldots)$, determines the axial shear stress perturbation τ_1 and a similar integral yields the cross-plane shear stress τ_2, where τ_1, τ_2 are the values of $\partial \tilde{U}/\partial N, \partial \tilde{W}/\partial N$ at the wall $N = 0$ respectively.

8.   Outlook and Exercises

This internal-flow study has been almost entirely on constrictions and/or branchings. The aim has been to seek out relatively simple configurations and basic properties first, as occurrence of important linear and nonlinear effects implies that mathematical accounts are often likely to be insightful. Assumptions made include those of mostly steady laminar incompressible-fluid motions over medium-to-large ranges of the Reynolds number; but see exercises below concerning unsteadiness, turbulence, etc. We believe the study may be of interest in terms of mathematical issues, real applications, the science of fluid dynamics and the clear interaction with direct numerical simulations.

In reality, there are often very complex networks to deal with in the practical situations of concern. Reconnections in addition to branchings are of interest whether predominantly inviscid or viscous-inviscid in nature. Flexible walls also arise in reality (see exercises) and, among other possibilities, they lead to integral equations stemming from (50) for instance. Again, the regime of small Reynolds numbers also has much fascination. The current contribution is further related to continuing work on many-body problems motivated by industrial as well as biomedical applications.

There are many other follow-on studies which could be undertaken. With that in mind, we refer to questions set elsewhere[1] and also to the following exercises E1–E10.

E1. Take the boundary layer system with no slip at $y = 0$, for steady flow; differentiate the momentum equation twice with respect to y and set $y = 0$; show that formally a square-root singularity can arise in the WSS as the latter tends to 0+ at some station $x = x_0$. Give a more detailed study based on the coordinate $y/(x_0 - x)^{1/4}$ to support the square-root finding. (This is the so-called Goldstein singularity. Removal of the singularity and marginal separation are related subsequent research areas.)

E2. Follow through with the FT working in Sec. 7.1 concerning a 3D constriction. Confirm the existence of a pole at $\alpha = -i\beta$. Find a formula for the streamwise WSS perturbation.

E3. Derive the integral equation (52) in detail, concerning multi-branching.

E4. Derive an integro-differential equation from $(e_1\partial_x^4 + e_2\partial_x^2 + e_3 + e_4\partial_T^2 + e_5\partial_T)F = \tilde{P} - \tilde{P}_0$ and (29) for interaction between condensed flow (Secs. 3 and 6) and a flexible wall, where e_1–e_5, $\tilde{P}_0$ are constants. Find the steady solution formally by means of a Laplace transform (LT) in x.

E5. For the case of 1-to-N branching in Sec. 4.2 deduce the behaviour in the limit of large N.

E6. For 3D motion through a constricted tube (Sec. 7.1), find the flow structure and solution if the incident velocity profile is not symmetric in the azimuthal direction, i.e., it depends on θ.

E7. Investigate the case corresponding to severe constriction for a non-symmetric channel (Sec. 6.1), and apply this to branching flows if possible.

E8. Examine the unsteady non-symmetric-channel and external triple-deck problems and their significance for linear and nonlinear viscous-inviscid waves and finite-time breakup.

E9. Add (compare (91)) a turbulent modelled term to the boundary layer system and follow the consequences of increasing that term, both in a classical boundary layer and in an interactive one.

E10. Examine and compare the influences of geometrical shape such as divergence angles, thickness, camber, positioning in 2D and 3D branchings as discussed in Secs. 4.1 and 7.2.

Model answer for E1: take the scaled form $u = \psi_Y, v = -\psi_x, uu_x + vu_Y = -p'(x) + u_{YY}$. Differentiating in Y gives

$$(\text{since } u_x + v_Y = 0) \quad uu_{xY} + vu_{YY} = u_{YYY} \qquad (120)$$

and so differentiating again we have

$$uu_{xYY} + u_Y u_{xY} + vu_{YYY} + v_Y u_{YY} = u_{YYYY}. \qquad (121)$$

Put $Y = 0$, given that $u = v = 0$ there, so $v_Y = 0$ also, and we obtain

$$\tau(x)\tau'(x) = -B(x), \qquad (122)$$

where at $Y = 0, \tau \equiv u_Y$ is the WSS and $B \equiv -u_{YYYY}$ is constant (>0) usually as x approaches x_0. Hence locally $\frac{1}{2}\tau^2 = -(x-x_0)B(x_0)$, since $\tau \to 0+$ at $x = x_0-$, and hence

$$\tau \sim [2(x_0 - x)B(x_0)]^{\frac{1}{2}} \text{ as } x \to x_0 - . \qquad (123)$$

This is the Goldstein singularity. In more detail, the boundary layer develops a sublayer in which $\eta \equiv Y(x_0 - x)^{-1/4}$ is $O(1)$ (similarly to (12)) and

$$\psi = \frac{1}{6}\lambda_0\eta^3(x_0 - x)^{\frac{3}{4}} + A\eta^2(x_0 - x) + \cdots, \qquad (124)$$

as $x \to x_0-$, where $\lambda_0(>0)$ is from the zero-WSS velocity profile ($u_0 \sim \frac{1}{2}\lambda_0 Y^2$) exactly at $x = x_0$ and $A = (B(x_0)/2)^{1/2}$ acts as a displacement effect. Outside of the sublayer, the majority of the flow where $Y \sim 1$ undergoes the displacement

$$\psi = \psi_0(Y) + \cdots, u = u_0(Y) + \left(\frac{2A}{\lambda_0}\right)(x_0 - x)^{\frac{1}{2}}u_0'(Y) + \cdots.$$
$$(125)$$

The boundary layer thickness $\delta(x)$ thus behaves as $(x_0-x)^{1/2}$ locally. That is, for a classical flow, where $p(x)$ is prescribed, whereas for an interactive flow $\delta(x)$ is prescribed (or is related to $p(x)$), which demands $A = 0$ and so then the singularity is absent. See papers by S. Goldstein in 1948, K. Stewartson in 1970 for fuller accounts.

Model answer for E2: the FT for 3D constriction gives the form, in a cleaner notation,

$$i\alpha U + V_Y + i\beta W = 0, \quad i\alpha Y(U,W) + (V,0) = -i(\alpha,\beta)P + (U,W)_{YY} \tag{126}$$

with requirements (109). Therefore, W satisfies

$$W_{\xi\xi} - \xi W = C, \text{ with } W = 0 \text{ at } Y = 0, \infty, \tag{127}$$

where $C \equiv i\beta(i\alpha)^{-\frac{2}{3}}P$. The solution is

$$W = C\mathcal{L}(\xi) \text{ with } \mathcal{L}(\xi) \equiv Ai(\xi) \int_0^\xi Ai^{-2}(\hat{\xi}) \left\{ \int_\infty^{\hat{\xi}} Ai(\bar{\xi})d\bar{\xi} \right\} d\hat{\xi}. \tag{128}$$

So (126) then requires $\tau_{\xi\xi} - \xi\tau = i\beta(i\alpha)^{-2/3}W$ to be solved for shear $\tau \equiv U_Y$ giving

$$\tau = -i\beta(i\alpha)^{-\frac{2}{3}}C\mathcal{L}'(\xi) + BAi(\xi). \tag{129}$$

Next, the conditions at $Y = 0, \infty$ are treated as per (26, 27) and yield the pressure and WSS solutions; in particular the P transform is

$$P = -3Ai'(0)(i\alpha)^{\frac{5}{3}} \frac{F}{(\alpha^2 + \beta^2)}. \tag{130}$$

This agrees with (28) for symmetric 3D cases where $\beta = 0$. For non-symmetric 3D cases however the Z-dependence or β effect induces a pole at $\alpha = -i\beta$ in (130).

Model answer for E4: eliminating $\tilde{P}$ yields the required equation,

$$e_1 F_{xxxx} + e_2 F_{xx} + e_3 F + e_4 F_{TT} + e_5 F_T = -\gamma \int_0^x F(s)(x-s)^{-\frac{2}{3}}ds - \tilde{P}_0 \tag{131}$$

for F in steady flow; (131) is to be solved subject to $F = F_x = 0$ at the ends $x = 0, x_1$, of a flexible patch. One way is to apply the LT with $x \to q$, which converts (131) to

$$F^l = \frac{\left[e_1(a + qb) - \tilde{P}_0 q^{-1} \right]}{\left[e_1 q^4 + e_2 q^2 + e_3 + \gamma\Gamma\left(\frac{1}{3}\right)q^{-\frac{1}{3}} \right]}. \tag{132}$$

The Bromwich inversion integral then yields F explicitly, aided by contour integration in the q-plane, and the two unknowns $a \equiv F_{xxx}$,

$b \equiv F_{xx}$ at $x = 0$ are determined by the two conditions at $x = x_1$. See papers by P. Carpenter *et al.* in 1985, 1986, J. Guneratne and T. Pedley in 2006, L. Pruessner and F. Smith in 2015.

Acknowledgments

Detailed comments by research students Samire Balta and Pietro Servini and by Dr Sergei Timoshin are acknowledged gratefully.

References

1. F. T. Smith, On internal fluid dynamics. *Bull. Math. Sci.* **2**, 125–180 (2012).
2. T. J. Pedley, Mathematical modelling of arterial fluid dynamics. *J. Eng. Math.* **47**, 419–444 (2003).
3. R. I. Sykes, On three-dimensional boundary layer flow over surface irregularities. *Proc. R. Soc. A* **373**, 311–329 (1980).
4. F. T. Smith, Properties, and a finite difference approach, for interactive three dimensional boundary layers. *UTRC* 83–46 (1983); see also Steady and unsteady boundary layer separation. *Ann. Rev. Fluid Mech.* **18**, 197–220 (1986).
5. F. T. Smith, Upstream interactions in channel flows. *J. Fluid Mech.* **79**, 631–655 (1977).
6. F. T. Smith and M. A. Jones, One-to-few and one-to-many branching tube flows. *J. Fluid Mech.* **423**, 1–31 (2000).
7. F. T. Smith, N. C. Ovenden, P. T. Franke and D. J. Doorly, What happens to pressure when a flow enters a side branch? *J. Fluid Mech.* **479**, 231–258 (2003).
8. I. J. Sobey, *Introduction to Interactive Boundary Layer Theory.* Oxford University Press, Oxford (2001).
9. L. D. Landau and E. M. Lifschitz, *Fluid Mechanics.* Reprinted 2000, 2nd edn. (2000).
10. F. T. Smith, The separating flow through a severely constricted symmetric tube. *J. Fluid Mech.* **90**, 725–754 (1979).
11. J. K. Comer, C. Kleinstreuer and Z. Zhang, Flow structures and particle deposition patterns in double-bifurcation airway models. Part 1. Air flow fields. *J. Fluid Mech.* **435**, 25–54 (2001).
12. F. T. Smith and M. A. Jones, AVM modelling by multi-branching tube flow: large flow rates and dual solutions. *Math. Med. and Biol.* **20**, 183–204 (2003).
13. P. W. Duck and O. R. Burggraf, Spectral solutions for three-dimensional triple-deck flow over surface topography. *J. Fluid Mech.* **162**, 1–22 (1986).
14. F. T. Smith, Flow through constricted or dilated pipes and channels. Parts 1, 2. *Quart. Mech. Appl. Math.* **29**, (1976).
15. S. N. Timoshin, Chapter in *Fluid and Solid Mechanics.* World Scientific Press, Singapore (2016).

Chapter 6

Fundamentals of Physiological Solid Mechanics

N. C. Ovenden* and C. L. Walsh

*Department of Mathematics, University College London,
Gower St, London WC1E 6BT, UK*

*Centre for Mathematics, Physics and Engineering in the
Life Sciences and Experimental Biology (CoMPLEX),
University College London, Gower St,
London WC1E 6BT, UK
Email: n.ovenden@ucl.ac.uk.*

This chapter introduces the fundamental concepts of elasticity and presents the application of continuum mechanics to modelling biological soft tissue and bone. The microstructure of tissue is examined and commonly-used constitutive relations presented. The chapter then continues by examining wave propagation in biological tissue, as well as providing further details pertaining to ultrasound imaging techniques. Finally, extensions to standard elasticity models that incorporate viscoelastic effects and the coupling between a porous solid and an interstitial fluid are briefly presented.

1. Introduction

Theories governing the behaviour of solids under the action of forces have found applications in a wide variety of mechanical and civil engineering problems over many decades. In recent times, such theories have been applied to understand the behaviour of biological soft tissue and bone structures, whether it is for interpreting medical images, modelling pathologies such as tumours, or understanding the complex fluid-structure interactions in the cardiovascular system. This chapter presents some introductory concepts in solid mechanics using an approach more slanted towards physiological modelling, as opposed to the traditional focus on industrial engineering applications.

The theory of elasticity is presented in Sec. 2, by first introducing the concepts of strain and stress and, from these, developing governing equations for conservation of momentum and energy. Both linear and nonlinear theories are explored and general forms of constitutive relations are given for an isotropic material, including hyperelastic materials possessing a strain energy density. Section 3 starts by examining the physical basis of common parameters used in the constitutive relations. This is followed by a discussion of the characteristic microstructure of tissues and some simplifying assumptions typically made in computational models. A brief review of constitutive relations commonly used to model biological tissue is then presented at the end of the section.

Medical imaging requires a detailed understanding of how waves propagate in human tissue and Sec. 4 explores the propagation of mechanical waves in tissue and why ultrasound imaging works as an effective tool for determining the internal structure of the body. The final two sections then briefly introduce two significant extensions to classical elasticity theory: (i) incorporating viscoelastic effects through strain-rate dependence of the material (Sec. 5) and (ii) poroelastic behaviour where the tissue is modelled more accurately as a porous elastic matrix containing an interstitial fluid (Sec. 6). Some exercises for the reader and suitable references to explore further are given at the end of the chapter.

2. Elasticity

The theory of elasticity is concerned with the reversible process of a solid body being deformed from an unstressed reference state to a stressed state by mechanical or thermal loads — with the assumption that removal of the loads leads to the body reassuming its unstressed reference state.

2.1. *Strain*

Suppose that the deformation can be represented by a displacement function $u(X)$, where every point in the unstressed body X is displaced to position $x = X + u(X)$ in the stressed configuration. Now, let's take two arbitrary neighbouring points in a volume undergoing such deformation, X and $X + \delta X$ (Fig. 1), and measure how the distance between those two points changes between the unstressed and stressed configurations. In the stressed configuration, the line element joining the two points, δx, is given by

$$\delta x = \delta X + u(X + \delta X) - u(X) = \delta X + (\delta X \cdot \nabla) u(X). \quad (2.1)$$

In the component form, where $X = (X_1, X_2, X_3)^T = X_i$ and $u = u_i$, it is then straightforward to show that

$$|\delta x|^2 - |\delta X|^2 = 2\mathcal{E}_{ij} \delta X_i \delta X_j, \quad (2.2)$$

where Einstein's summation convention has been used and

$$\mathcal{E}_{ij} = \frac{1}{2} \left(\frac{\partial u_i}{\partial X_j} + \frac{\partial u_j}{\partial X_i} + \frac{\partial u_k}{\partial X_i} \frac{\partial u_k}{\partial X_j} \right). \quad (2.3)$$

The fact that the length is unchanged only when $\mathcal{E}_{ij} = 0$ suggests that $\mathcal{E}_{ij}$ is a measure of the strain of the material. Furthermore, $\mathcal{E}_{ij}$

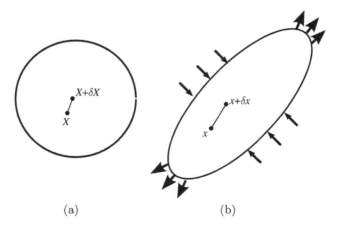

(a)   (b)

Fig. 1.   (a) An elastic body and a line element δX in the unstressed reference state; (b) the deformation of the same body and the line element under the action of external forces.

obeys the necessary transformation laws of a second-rank Cartesian tensor, meaning that if the coordinate axes are rotated such that the reference location and displacement transform as $X'_i = R_{ij}X_j$ and $u'_i = R_{ij}u_j$, for a given orthogonal matrix $\boldsymbol{R} = R_{ij}$ (where by definition $\boldsymbol{R}^{-1} = \boldsymbol{R}^T$), then the strain tensor transforms[a] as

$$\mathcal{E}'_{ij} = R_{ip}R_{jq}\mathcal{E}_{pq}. \tag{2.4}$$

For this reason, $\mathcal{E}_{ij}$ is defined as the *strain tensor* and it is straightforward to show that it possesses the important physical property of invariance to both a uniform translation of the entire body and rigid body rotation.

A more general approach to measuring strain is to define a function $\boldsymbol{x}(\boldsymbol{X}, t)$ (where t is time) that maps a solid particle at point $\boldsymbol{X}$ in the reference state to its displaced position in the stressed state. The deformation of a line element is then given by $\mathrm{d}x_i = F_{ij}\mathrm{d}X_j$, where

$$F_{ij} = \frac{\partial x_i}{\partial X_j}, \tag{2.5}$$

is called the *deformation gradient*. By determining once again the change in the length squared of a line element as above, we find that

$$|\mathrm{d}\boldsymbol{x}|^2 = F_{ki}F_{kj}\mathrm{d}X_i\mathrm{d}X_j. \tag{2.6}$$

$\mathcal{C}_{ij} = F_{ki}F_{kj}$ is known as the *right Cauchy–Green deformation tensor* and its relationship with the strain tensor defined above can be simply expressed (by using the Kronecker delta δ_{ij}) by

$$\mathcal{E}_{ij} = \frac{1}{2}\left(\mathcal{C}_{ij} - \delta_{ij}\right) \quad \text{and} \quad \mathcal{C}_{ij} = \delta_{ij} + 2\mathcal{E}_{ij}. \tag{2.7}$$

Clearly, from this relationship, $\mathcal{C}_{ij}$ also obeys the necessary transformation laws of a second-rank tensor.

---

[a]While distinction between covariant and contravariant indices is necessary for general tensors, the two are equivalent for tensors in 3D Euclidean space; such tensors are known as Cartesian tensors — see Ref. 1.

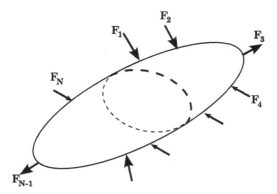

Fig. 2.   A solid body in equilibrium under the action of N external forces.

2.2.   *Stress*

Imagine a body in equilibrium under the action of N external forces: $\boldsymbol{F}_1$, $\boldsymbol{F}_2$, ..., $\boldsymbol{F}_N$ as shown in Fig. 2. If the body is divided into two parts (as illustrated by the dotted line) then the adjacent interior surfaces would each apply a continuous, equal and opposite stress or traction (force per unit area) to the other. In general, such stress is likely to have a tensile or compressive component normal to the surface, as well as a shearing component that acts in the plane of the surface.

To analyse this further, take a small cuboid element at an interior point in the body, the sides of which lie parallel to some imposed Cartesian coordinate axes (x_1, x_2, x_3), and let the cuboid have dimensions δx_1, δx_2 and δx_3. Figure 3 shows that the stress on each side of the cuboid can be split into three components in the form $\sigma_{ij}n_j$ where n_j is the normal to the surface. Hence, σ_{11}, σ_{22} and σ_{33} are stresses acting normal to the respective sides of the cuboid, with positive values implying a tension (or stretch) and negative values implying a compression. The other stress components σ_{ij} for $i \neq j$ are shear stresses which act in the plane of the cuboid surface in question. By considering the angular momentum of the cuboid we find that these shear components cannot all be independent. For instance, taking moments about the x_3 axis and assuming conservation of angular

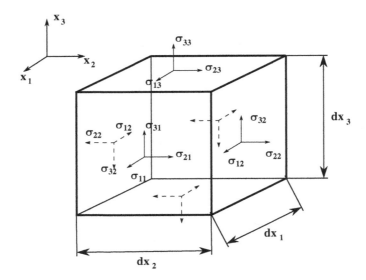

Fig. 3. The stress components acting on the faces of a small cuboid element.

momentum yields

$$2\sigma_{21}\left(\frac{\delta x}{2}\right)\delta x_2\delta x_3 - 2\sigma_{12}\left(\frac{\delta x_2}{2}\right)\delta x_1\delta x_3 = 0; \qquad (2.8)$$

similarly, two equations of the same form can be obtained from taking conservation of angular momentum about the x_1 and x_2 axes. As a result, we find a necessary symmetry condition for the stress tensor that

$$\sigma_{ij} = \sigma_{ji}. \qquad (2.9)$$

Hence, in the limit $\delta x_i \to 0$ for $i = 1, 2, 3$, only six stress components (σ_{11}, σ_{22}, σ_{33}, σ_{12}, σ_{13}, σ_{23}), as opposed to nine, are required to specify the stress field at any point in the solid body.

A governing equation for the elasticity of a solid body can be obtained by applying Newton's second law to some *arbitrary* material volume $V(t)$ that deforms under the action of the applied forces. We assume that the rate of change of momentum of the material in $V(t)$ equals the forces applied to the material volume. These forces can be of two basic types: (i) *body forces*, such as those imposed externally by gravity or an electromagnetic field, which will be represented here by the vector $\boldsymbol{g} = g_i$, and (ii) *surface forces* that are stresses applied to the external boundary of the material volume, $\partial V(t)$. Conservation of momentum can thus be expressed as follows:

$$\frac{\mathrm{d}}{\mathrm{d}t} \iiint\limits_{V(t)} \frac{\partial u_i}{\partial t} \rho \, \mathrm{d}\boldsymbol{x} = \iiint\limits_{V(t)} g_i \rho \, \mathrm{d}\boldsymbol{x} + \iint\limits_{\partial V(t)} \sigma_{ij} n_j \, \mathrm{d}a, \qquad (2.10)$$

where $\mathrm{d}a$ is an infinitesimal area on the boundary of the material volume. If the density in the reference state is given by $\rho_0(\boldsymbol{X})$, we can bring the time derivative into the integral on the left-hand side by observing that $\rho(\boldsymbol{x}) \, \mathrm{d}\boldsymbol{x} = \rho_0(\boldsymbol{X}) \, \mathrm{d}\boldsymbol{X}$ and is therefore independent of time. Then, by assuming the integrands to be continuous and by applying the divergence theorem, we thus arrive at *Cauchy's momentum equation*,

$$\rho \frac{\partial^2 u_i}{\partial t^2} = \rho g_i + \frac{\partial \sigma_{ij}}{\partial x_j} \quad \text{or} \quad \rho \frac{\partial^2 \boldsymbol{u}}{\partial t^2} = \rho \boldsymbol{g} + (\nabla \cdot \boldsymbol{\sigma}). \qquad (2.11)$$

This simple equation is remarkably universal and can indeed be applied to any continuous medium by establishing a constitutive relationship between the stress field σ_{ij} and the displacement u_i. Unfortunately, a troublesome difficulty arises at this stage in that (2.11) contains variables, like density, and derivatives (e.g., $\partial/\partial x_i$) which are typically defined in respect to the (Eulerian) coordinates of the (*unknown*) stressed state $\boldsymbol{x}$, whereas other variables present, such as displacement $u_i(\boldsymbol{X})$ and thus strain $\mathcal{E}_{ij}$, are typically defined in respect to the reference configuration $\boldsymbol{X}$ (Lagrangian coordinates). This difficulty can be overcome by either (i) assuming small (infinitesimal) displacements, leading to a governing equation for linear elasticity or (ii) finding a stress tensor in Lagrangian coordinates and from there a constitutive relation. Both of these approaches are presented in the subsections below.

2.3. *Linear elasticity*

For linear elasticity, it is assumed that the displacements $\boldsymbol{u}$ are small relative to the typical length scale(s) of the problem and, as a consequence, the Eulerian and Lagrangian coordinates, $\boldsymbol{x}$ and $\boldsymbol{X}$, are equal to leading order in powers of $|\boldsymbol{u}|$. Moreover, as $\rho \mathrm{d}\boldsymbol{x} = \rho_0 \mathrm{d}\boldsymbol{X}$

the changes in density between stressed and reference states can be neglected. Thus, one can approximate $\partial\sigma_{ij}/\partial x_j$ by $\partial\sigma_{ij}/\partial X_j$ and $\rho(\boldsymbol{x})$ by $\rho_0(\boldsymbol{X})$ in Eq. (2.11). The deformation gradient F_{ij} and right Cauchy–Green deformation tensor $\mathcal{C}_{ij}$ are equal to the identity matrix to leading order and the quadratic term in the strain tensor (2.3) is negligible to leading order leading to a linearised strain tensor of the form

$$\mathcal{E}_{ij} \approx \varepsilon_{ij} = \frac{1}{2}\left(\frac{\partial u_i}{\partial X_j} + \frac{\partial u_j}{\partial X_i}\right). \tag{2.12}$$

Linear elasticity restricts our constitutive relation to a linear relation between stress and strain such as in Hooke's law. The most general form of any such constitutive relation is

$$\sigma_{ij} = \mathcal{A}_{ijkl}\varepsilon_{kl}, \tag{2.13}$$

where $\mathcal{A}$ is a fourth rank tensor that can be transformed to a new coordinate system via the relation

$$\mathcal{A}'_{ijkl} = L_{ip}L_{jq}L_{kr}L_{ls}\mathcal{A}_{pqrs} \tag{2.14}$$

for any orthogonal matrix $\boldsymbol{L}$, where $\boldsymbol{L}^T = \boldsymbol{L}^{-1}$. Similarly, an inverse relationship holds via another fourth rank tensor with similar properties known as the compliance tensor b, such that $\varepsilon_{ij} = b_{ijkl}\sigma_{kl}$. Both $\mathcal{A}_{ijkl}$ and b_{ijkl} are required to satisfy the already identified symmetry properties of the stress and strain tensors, i.e., that $\sigma_{ij} = \sigma_{ji}$ and $\varepsilon_{ij} = \varepsilon_{ji}$. For this reason,

$$\mathcal{A}_{ijkl} = \mathcal{A}_{ijlk} = \mathcal{A}_{jikl}, \tag{2.15}$$

reducing the degrees of freedom significantly. Further simplification is possible for a material that is *isotropic* so that $\mathcal{A}_{ijkl}$ is invariant under all rotations of the coordinate axes as in (2.14). One might expect in this case that Hooke's law $\sigma_{ij} = E\varepsilon_{ij}$ is at least permitted, where E is Young's modulus. In fact, it is possible to show (see Ref. 2, pp. 7–9) that a two-parameter family of constitutive relations of the form

$$\mathcal{A}_{ijkl} = \lambda\delta_{ij}\delta_{kl} + \mu\left(\delta_{ik}\delta_{jl} + \delta_{jk}\delta_{il}\right), \tag{2.16}$$

offers the only permitted set of linear constitutive relations for an isotropic material. The two parameters λ and μ are known as Lamé's parameters and μ is also called the *shear modulus*. Substituting (2.16) into (2.13) leads to a stress–strain relation of the form,

$$\sigma_{ij} = \lambda\varepsilon_{kk}\delta_{ij} + 2\mu\varepsilon_{ij} \qquad (2.17)$$

and substituting this into Cauchy's equation (2.11) with the approximations $x \to X$ and $\rho \to \rho_0$ yields the following equation for the displacement field:

$$\rho_0\frac{\partial^2 u_i}{\partial t^2} = \rho_0 g_i + (\lambda + \mu)\frac{\partial^2 u_j}{\partial x_i \partial x_j} + \mu\frac{\partial^2 u_i}{\partial x_j{}^2} \qquad (2.18)$$

which is known as the *Navier equation*. A closed system of three equations for the three unknown components of the displacement field u_i has thus been derived. Three boundary conditions are required at every point on the boundary of the domain. These boundary conditions can either be a prescribed traction/stress $\sigma_{ij}n_j$, where n_j is the normal to the boundary surface, or the prescribed displacement field at the boundary u_i. Prescription of both displacement and traction at any point on the boundary is likely to lead to an ill-posed problem.

2.4. *Nonlinear elasticity*

The Cauchy stress σ_{ij} is regarded as the so-called *true stress* because it can be measured instantaneously on the deformed body during experiments. However, for modelling purposes a Lagrangian stress tensor expressed in terms of the reference state coordinates X is preferable as it can be directly coupled to the strain tensor (expressed in X) in a straightforward manner to derive a constitutive relation. To determine a Lagrangian stress tensor, we must return to Fig. 1 and the Cauchy equation given by (2.11). By considering an arbitrary volume in the reference frame V_0 and its deformed volume in the stressed state $V(t)$, conservation of mass implies that

$$\iiint\limits_{V_0} \rho_0(X)\mathrm{d}V = \iiint\limits_{V(t)} \rho(x,t)\mathrm{d}v = \iiint\limits_{V_0} \rho(X,t)\det(F)\mathrm{d}V, \qquad (2.19)$$

as the infinitesimal volumes in reference and stressed state are related by $\mathrm{d}v = \det(F)\mathrm{d}V$. Given that the volume is arbitrary we must therefore conclude that

$$\rho_0 = \det(F)\rho. \tag{2.20}$$

Now, suppose there exists a Lagrangian stress tensor $T = T_{ij}$. If we take the Lagrangian stress acting on an area in the reference configuration A_0 that is transformed into a Cauchy stress acting on an area $A(t)$ in the stressed state then, for the overall force to be equal in both configurations, it must be true that

$$\iint\limits_{A_0} T \cdot \mathrm{d}\boldsymbol{A} = \iint\limits_{A(t)} \sigma \cdot \mathrm{d}\boldsymbol{a}. \tag{2.21}$$

To determine exactly how $\mathrm{d}\boldsymbol{a}$ and $\mathrm{d}\boldsymbol{A}$ are related, consider the infinitesimal volume $\mathrm{d}\boldsymbol{X} \cdot \mathrm{d}\boldsymbol{A}$ that transforms to $\mathrm{d}\boldsymbol{x} \cdot \mathrm{d}\boldsymbol{a}$ in the stressed state. From our knowledge of infinitesimal volume changes

$$\mathrm{d}\boldsymbol{x} \cdot \mathrm{d}\boldsymbol{a} = \det(F)\mathrm{d}\boldsymbol{X} \cdot \mathrm{d}\boldsymbol{A} \tag{2.22}$$

and, by the definition of the deformation gradient F_{ij}, we also know that

$$\mathrm{d}\boldsymbol{x} \cdot \mathrm{d}\boldsymbol{a} = \mathrm{d}\boldsymbol{X} \cdot (F^T \mathrm{d}\boldsymbol{a}). \tag{2.23}$$

Hence, equating these two expression yields

$$\mathrm{d}\boldsymbol{a} = \det(F) \left(F^T\right)^{-1} \mathrm{d}\boldsymbol{A} \tag{2.24}$$

and, on returning to the force equation (2.21) above, we see that this equation can only be satisfied for any arbitrary area if the Lagrangian stress tensor is of the form

$$T_{ij} = \det(F)\,\sigma_{ik}\,F_{jk}^{-1}. \tag{2.25}$$

T_{ij} is a second-rank tensor known as the *first Piola–Kirchhoff Stress Tensor*. One immediate observation of the Lagrangian stress tensor is that, unlike the Cauchy Stress σ_{ij}, it is not symmetric, $T_{ij} \neq T_{ji}$. The symmetry can be restored, however, by introducing another Lagrangian tensor, known as the *second Piola–Kirchhoff Stress Tensor*, which is defined thus

$$S_{ij} = F_{ik}^{-1}T_{kj} = \det(F)F_{ik}^{-1}\sigma_{kp}F_{jp}^{-1} \tag{2.26}$$

It is straightforward to show that S_{ij} is symmetric.

Conservation of momentum can now be expressed fully in Lagrangian variables by substituting (2.20) and (2.25) into the Cauchy equation (2.11) to obtain,

$$\rho_0 \frac{\partial^2 x_i}{\partial t^2} = \rho_0 g_i + \det(F) F_{kj}^{-1} \frac{\partial}{\partial X_k} \left(\frac{1}{\det(F)} T_{il} F_{jl} \right). \qquad (2.27)$$

Subsequently, by using the relation

$$\frac{\partial F_{jk}}{\partial x_j} = \frac{1}{\det(F)} \frac{\partial}{\partial X_k} \det(F), \qquad (2.28)$$

(which can be proven in an arbitrary volume via the divergence theorem) the Lagrangian form of Cauchy's equation for nonlinear elasticity can be derived as

$$\rho_0 \frac{\partial^2 x_i}{\partial t^2} = \rho_0 g_i + \frac{\partial T_{ij}}{\partial X_j}. \qquad (2.29)$$

Of course, closing this model still requires a constitutive relation relating the Lagrangian stress to the strain, which is itself determined from the deformation gradient, thus $T_{ij} = T_{ij}(F_{kl})$. It is sensible to ask what constraints are required for such a constitutive relation to be physical? To determine these constraints, we must now examine energy conservation.

2.5. *Energy conservation*

An energy equation can be derived by multiplying the Lagrangian form of Cauchy's equation (2.29) by the local velocity field $\partial x_i/\partial t$ and then integrating over the volume of a solid body in its reference state V_0, yielding

$$\iiint\limits_{V_0} \rho_0 \frac{\partial x_i}{\partial t} \frac{\partial^2 x_i}{\partial t^2} \mathrm{d}\boldsymbol{X} = \iiint\limits_{V_0} \left[\rho_0 g_i + \frac{\partial T_{ij}}{\partial X_j} \right] \frac{\partial x_i}{\partial t} \mathrm{d}\boldsymbol{X}. \qquad (2.30)$$

Via some further manipulation and use of the divergence theorem, a relationship can be obtained between the rate of change of kinetic energy in the body, the potential (stored) energy in the elastic body, and the work done by body forces and external stresses, thus

$$\frac{\mathrm{d}}{\mathrm{d}t}\iiint_{V_0} \frac{\rho_0}{2}\left(\frac{\partial x_i}{\partial t}\right)^2 \mathrm{d}\boldsymbol{X} + \iiint_{V_0} T_{ij}\frac{\partial F_{ij}}{\partial t}\mathrm{d}\boldsymbol{X}$$

$$= \iiint_{V_0} \rho_0 g_i \frac{\partial x_i}{\partial t}\mathrm{d}\boldsymbol{X} + \iint_{\partial V_0} \frac{\partial x_i}{\partial t}T_{ij}n_j\mathrm{d}A,$$

$$(2.31)$$

where n_j is the normal to the body surface in reference coordinates. An important physical constraint on the validity of a constitutive relation must be that, when a body is deformed from one state to another and then back again, there should be *no* net energy gain. A sufficient condition to ensure this is to make the strain energy stored *path-independent*; in other words, dependent only on the initial and final states of F_{ij} and not on the way (path) such a deformation was achieved. This implies the existence, therefore, of a scalar strain energy density $\mathcal{W}(F_{ij})$, where

$$T_{ij} = \frac{\partial \mathcal{W}}{\partial F_{ij}}. \tag{2.32}$$

The existence of such a strain energy density allows us to write the potential energy of the elastic body in the form

$$\iiint_{V_0} T_{ij}\frac{\partial F_{ij}}{\partial t}\mathrm{d}\boldsymbol{X} = \frac{\mathrm{d}}{\mathrm{d}t}\iiint_{V_0} \mathcal{W}(F_{ij})\,\mathrm{d}\boldsymbol{X} \tag{2.33}$$

which must integrate to zero over any periodic loading cycle. Materials that possess a strain energy density $\mathcal{W}(F_{ij})$ are referred to as *hyperelastic* and represent by far the most commonly used constitutive relations in solid mechanics. Here, we limit our attention to isotropic hyperelastic materials and discuss the possible forms of constitutive relation. It seems intuitively clear that a strain energy density function must possess two important properties:

(i) $\mathcal{W}(F_{ij})$ should have a minimum (that can be set at zero without loss of generality) when $F_{ij} = \delta_{ij}$; and

(ii) $\mathcal{W}$ should be invariant under rigid body rotations and uniform translations of the reference state.

It is possible to write the (presumably) non-singular deformation gradient F_{ij} as the product of an orthogonal matrix $\boldsymbol{R}$, representing a rigid body rotation, and a positive symmetric definite matrix $\boldsymbol{U}$ with real positive eigenvalues α_i (Polar Decomposition). The positive symmetric matrix $\boldsymbol{U}$ can, in turn, be diagonalised in the form $\boldsymbol{U} = \boldsymbol{P}^T \Lambda \boldsymbol{P}$ where $\boldsymbol{P}$ is another orthogonal matrix and

$$\Lambda = \begin{pmatrix} \alpha_1 & 0 & 0 \\ 0 & \alpha_2 & 0 \\ 0 & 0 & \alpha_3 \end{pmatrix}. \tag{2.34}$$

The eigenvalues α_i, $i = 1, 2, 3$ represent the so-called *principal stretches* in the *principal directions* determined by the eigenvectors. From the definition of the right Cauchy–Green deformation tensor, it is consequently the case that $\mathcal{C} = \boldsymbol{U}^2$ and thus its eigenvalues are α_i^2 (as $\mathcal{C} = P^T \Lambda^2 P$).

For any isotropic material, it must be the case that the strain energy density $\mathcal{W}$ is invariant under rigid rotations. This means that $\mathcal{W}$'s dependence on the deformation gradient F_{ij} can only be implicit through the symmetric right Cauchy–Green deformation tensor $\mathcal{C}_{ij}$. Indeed, for all orthogonal matrices $\boldsymbol{L}$, such invariance implies that

$$\mathcal{W}\left(\boldsymbol{L}\mathcal{C}\boldsymbol{L}^T\right) = \mathcal{W}(\mathcal{C}) \tag{2.35}$$

and, as $\mathcal{C}$ is itself diagonalised by some orthogonal matrix $\boldsymbol{P}$, then one must infer that the strain energy can only be a symmetric function of the eigenvalues of $\mathcal{C}$. Therefore,

$$\mathcal{W}(F_{ij}) = \mathcal{W}(\alpha_1^2, \alpha_2^2, \alpha_3^2). \tag{2.36}$$

However, as any rigid rotation of the axes permits the eigenvalues themselves to permutate, the strain energy must be further restricted to only depend on the *isotropic invariants* of the right Cauchy–Green deformation tensor, $\mathcal{I}_i$, which are the coefficients of the characteristic polynominal of $\mathcal{C}$:

$$\det\left(\xi\mathcal{I} - \mathcal{C}\right) = \xi^3 - \mathcal{I}_1(\mathcal{C})\xi^2 + \mathcal{I}_2(\mathcal{C})\xi - \mathcal{I}_3(\mathcal{C}) = 0. \tag{2.37}$$

The isotropic invariants are explicitly,

$$\mathcal{I}_1(\mathcal{C}) = \alpha_1^2 + \alpha_2^2 + \alpha_3^2 = \mathcal{C}_{kk},$$

$$\mathcal{I}_2(\mathcal{C}) = \alpha_1^2\alpha_2^2 + \alpha_1^2\alpha_3^2 + \alpha_2^2\alpha_3^3 = \frac{1}{2}\left[(\text{tr}(\mathcal{C}))^2 - \text{tr}\left(\mathcal{C}^2\right)\right],$$

$$\mathcal{I}_3(\mathcal{C}) = \det(\mathcal{C}) = (\alpha_1\alpha_2\alpha_3)^2. \tag{2.38}$$

As $\mathcal{C}_{ij} = F_{ki}F_{kj}$, the first and second Piola–Kirchhoff stress tensors for a hyperelastic material are related to the strain energy density function in the following way:

$$\mathcal{T}_{ij} = \frac{\partial \mathcal{W}}{\partial F_{ij}} = 2F_{ik}\frac{\partial \mathcal{W}}{\partial \mathcal{C}_{kj}} \quad \text{and} \quad \mathcal{S}_{ij} = F_{ik}^{-1}\mathcal{T}_{kj} = 2\frac{\partial \mathcal{W}}{\partial \mathcal{C}_{ij}}. \tag{2.39}$$

Given now that our strain energy must take the form

$$\mathcal{W} = \mathcal{W}\Big(\mathcal{I}_1(\mathcal{C}), \mathcal{I}_2(\mathcal{C}), \mathcal{I}_3(\mathcal{C})\Big) \tag{2.40}$$

and, noting that $\mathcal{W}$ should ideally be positive and have a minimum of zero for the case $\mathcal{I}_1 = 3$, $\mathcal{I}_2 = 3$, and $\mathcal{I}_3 = 1$, the second Piola–Kirchhoff stress tensor can be written thus,

$$\mathcal{S}_{ij} = 2\left(\frac{\partial \mathcal{W}}{\partial \mathcal{I}_1}\frac{\partial \mathcal{I}_1}{\partial \mathcal{C}_{ij}} + \frac{\partial \mathcal{W}}{\partial \mathcal{I}_2}\frac{\partial \mathcal{I}_2}{\partial \mathcal{C}_{ij}} + \frac{\partial \mathcal{W}}{\partial \mathcal{I}_3}\frac{\partial \mathcal{I}_3}{\partial \mathcal{C}_{ij}}\right) \tag{2.41}$$

Then, using the following relations (that can be derived from the explicit forms of the invariants):

$$\frac{\partial \mathcal{I}_1}{\partial \mathcal{C}_{ij}} = \delta_{ij}, \quad \frac{\partial \mathcal{I}_2}{\partial \mathcal{C}_{ij}} = \mathcal{I}_1\delta_{ij} - \mathcal{C}_{ij} \quad \text{and}$$

$$\frac{\partial \mathcal{I}_3}{\partial \mathcal{C}_{ij}} = \mathcal{I}_2\delta_{ij} - \mathcal{I}_1\mathcal{C}_{ij} + \mathcal{C}_{ik}\mathcal{C}_{kj} = \mathcal{I}_3\mathcal{C}_{ij}^{-1} \tag{2.42}$$

the following general form of the symmetric Lagrangian stress tensor can be obtained:

$$\mathcal{S}_{ij} = 2\left[\left(\frac{\partial \mathcal{W}}{\partial \mathcal{I}_1} + \mathcal{I}_1\frac{\partial \mathcal{W}}{\partial \mathcal{I}_2} + \mathcal{I}_2\frac{\partial \mathcal{W}}{\partial \mathcal{I}_3}\right)\delta_{ij} \right.$$

$$\left. -\left(\frac{\partial \mathcal{W}}{\partial \mathcal{I}_2} + \mathcal{I}_1\frac{\partial \mathcal{W}}{\partial \mathcal{I}_3}\right)\mathcal{C}_{ij} + \frac{\partial \mathcal{W}}{\partial \mathcal{I}_3}\mathcal{C}_{ik}\mathcal{C}_{kj}\right]. \tag{2.43}$$

The Cauchy stress σ_{ij} can also be determined from $\mathcal{W}$ via the relations (2.25) and (2.39) as

$$\sigma_{ij} = \frac{1}{\det(F_{ij})} \frac{\partial \mathcal{W}}{\partial F_{ik}} F_{jk}. \tag{2.44}$$

However, it can be shown that σ_{ij} is more elegantly expressed in terms of the so-called *left* Cauchy–Green deformation tensor, defined by $\mathcal{B}_{ij} = F_{ik}F_{jk}$, in the following manner:

$$\sigma_{ij} = \frac{1}{\sqrt{\mathcal{I}_3}} \left[2\mathcal{I}_3 \frac{\partial \mathcal{W}}{\partial \mathcal{I}_3} \delta_{ij} + 2 \left(\frac{\partial \mathcal{W}}{\partial \mathcal{I}_1} + \mathcal{I}_1 \frac{\partial \mathcal{W}}{\partial \mathcal{I}_2} \right) \mathcal{B}_{ij} - 2 \frac{\partial \mathcal{W}}{\partial \mathcal{I}_2} \mathcal{B}_{ik}\mathcal{B}_{kj} \right]. \tag{2.45}$$

Such elegance can be explained simply because, while the second Piola–Kirchoff tensor shares the same eigenvectors as $\mathcal{C}_{ij}$, the true (Cauchy) stress shares its principal axes with those of $\mathcal{B}_{ij}$.

Is the linear elasticity relation hyperelastic? It is useful perhaps to confirm a degree of consistency here with the theory of linear elasticity via the original strain tensor $\mathcal{E}_{ij}$. From the observed relationship seen earlier, (2.7), it is straightforward to show that

$$\frac{\partial \mathcal{W}}{\partial \mathcal{C}_{ij}} = \frac{1}{2} \frac{\partial \mathcal{W}}{\partial \mathcal{E}_{ij}}. \tag{2.46}$$

And, directly from the definition of linear stress (2.17), a strain energy function can therefore be determined of the form

$$\mathcal{W} = A_{ijkl}\varepsilon_{ij}\varepsilon_{kl} = \frac{1}{2}\lambda\delta_{ij}\delta_{kl}\varepsilon_{ij}\varepsilon_{kl} + \mu \left(\delta_{ik}\delta_{jl} + \delta_{il}\delta_{jk} \right) \varepsilon_{ij}\varepsilon_{kl}. \tag{2.47}$$

In the next section, constitutive relations to replicate the behaviour of biological soft tissues are presented. For completeness, we note that an isotropic material that is Cauchy elastic but *not* hyperelastic has a constitutive equation of the form

$$\mathcal{S}_{ij} = \mathcal{S}(\mathcal{C}), \tag{2.48}$$

but no strain energy function satisfying (2.32) exists. For problems involving hyperelastic materials, it is often easier numerically to find a displacment field that minimises

$$\iiint\limits_{V_0} \mathcal{W} \, \mathrm{d}\boldsymbol{X}, \tag{2.49}$$

than to numerically solve the nonlinear Cauchy equation in either Lagrangian (2.29) or Eulerian coordinates. For this reason, finite-element approaches are commonly used in solid mechanics problems as they are well suited to solving minimisation problems akin to (2.49).

3.   Constitutive Relations for Human Tissue

3.1.   *Material parameters*

Constitutive relations may be formulated either by relating the Lagrangian stress tensors ($\mathcal{T}_{ij}$ or $\mathcal{S}_{ij}$) to the strain or deformation tensors ($\mathcal{E}_{ij}$, $\mathcal{C}_{ij}$ or ε_{ij}) or in the form of a strain energy density function $\mathcal{W}(F)$. Formulating appropriate constitutive equations for biological tissues remains a difficult problem that much biomechanical research has been dedicated to over the past few decades. There are two approaches to developing a constitutive relation:

(i) A theoretical approach — based on knowledge of the microstructure of the material;

(ii) An experimental approach — based on curve fitting to data from specific experiments.

It is crucial to understand that constitutive equations only apply to a specific material under *specific conditions*. A relation that may be valid for tendon under normal walking conditions may not be suitable to explain a sudden tendon rupture generated by excessive stress; see Sharpe[3] for more discussion.

Tissues in general are nonlinear, and the most simple nonlinear constitutive relation is the Saint Venant–Kirchhoff relation. You will recognise its formulation from the linear constitutive equation given by Eq. (2.17) in the previous section,

$$\mathcal{S}_{ij} = \lambda \, \mathcal{E}_{kk}\delta_{ij} + 2\mu\mathcal{E}_{ij}, \tag{3.1}$$

but, here, with the Cauchy stress σ_{ij} replaced by the second Piola–Kirchhoff stress tensor S_{ij} and the linearised strain tensor ε_{ij} replaced by the strain tensor $\mathcal{E}_{ij}$. The parameters λ and μ have already been introduced as the Lamé parameters. These can also be rewritten in terms of three other commonly-used parameters, E, ν and K, given by the following relations:

$$\nu = \frac{\lambda}{2(\lambda + \mu)}, \quad \mu = \frac{E}{2(1 + \nu)}, \quad \lambda = \frac{\nu E}{(1 + \nu)(1 - 2\nu)},$$

$$E = \frac{\mu(3\lambda + 2\mu)}{\lambda + \mu} \quad \text{and} \quad K = \lambda + \frac{2}{3}\mu. \tag{3.2}$$

The physical meaning of these parameters is explained below.

Poisson's ratio ν

Poisson's ratio describes the tendency of materials to contract along the axes perpendicular to a force exerting an extension (see Fig. 4). Thus ν can be defined as the negative ratio of the relative contraction strain (or transverse strain) perpendicular to the applied load to the relative extension strain (or axial strain) in the direction of the applied load. Hence,

$$\nu = -\frac{de_{transverse}}{de_{axial}} = -\frac{de_y}{de_x} = -\frac{de_z}{de_x},$$

$$\text{where } de_x = \frac{dx}{x}, \quad de_y = \frac{dy}{y}, \quad de_z = \frac{dz}{z}.$$

Poisson ratio's has a limited range in reality, $-1 \leq \nu \leq 0.5$ where $\nu = 0.5$ denotes an incompressible material. Steel has $\nu \approx 0.3$ and

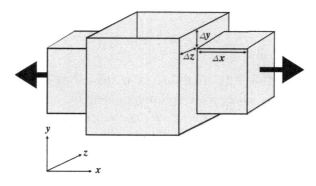

Fig. 4. Illustrating the Poisson ratio for a typical material ($\nu > 0$) where an extensional strain in the x direction leads to contractions in the directions perpendicular to the applied load.

rubbers typically approach the incompressibility limit of $\nu \approx 0.5$. Many types of biological tissues are also assumed to have $\nu \approx 0.5$. While most materials have a positive Poisson's ratio, certain materials, called *auxetic*, have $\nu < 0$ and counterintuitively expand transversely under a tensile strain. Common auxetic materials are cellular solids, including foams, and it has been shown that some biological tissues, including skin, bone and arterial walls, also exhibit auxetic behaviour.

Young's modulus E

Young's modulus is defined as the ratio of tensile stress to extensional strain. It is the constant of proportionality in Hooke's law or, equivalently, the gradient of a uniaxial stress strain curve. It has units of N/m^2 or Pa and takes very large values for stiff materials, such as steel, which has has $E = 2.10 \times 10^{11} \, N/m^2 = 210 GPa$. For tissues within the body, there is a wide range of E, from as high as 80GPa for bone, $1 - 2GPa$ for a tendon, to $1 - 2kPa$ for the liver.[4]

Shear modulus μ

The shear modulus describes a material's resistance to shear deformation or bending. It is defined as the ratio of shear stress to shear strain and has units Pa. The shear modulus also has a wide range of values across different tissue types, from $\mu \sim 1Pa$ for the eye to $\mu \sim 1 - 2MPa$ for cartilage.[4]

Bulk modulus K and λ

K is known as the bulk modulus as it describes the resistance of a material to any change in volume caused by an applied hydrostatic pressure (p); thus $p = -K\varepsilon_{kk}$. K is strictly positive and extremely large for almost all tissues, tending to infinity as the material approaches incompressibility. The Lamé parameter λ, on the other hand, is the only one of the five parameters to have no direct physical meaning. It is, however, often taken to be closely associated to the compressibility of the material as from Eqs. (3.2) one can see that for large λ, $K \approx \lambda$.

Parameters values and material behaviour

When modelling a particular system it is helpful to accurately estimate the values of these parameters for the material(s) in question. Often their relative sizes enable assumptions to be made regarding the form of constitutive equation. Naturally, in biological tissues, there are certainly other parameters to consider, such as those associated with more complex constitutive relations including viscoelastic and poroelastic models; these will be covered in later sections.

3.2. Structure of tissues

We now turn our attention to the microstructure of tissues. All tissues within the body are made up of a cellular and an acellular component. In soft tissues, the cellular component consists of cells that can exert some sort of contractile force, e.g., muscle cells or fibroblasts. These cells have a wide variety of effects on tissue behaviour and the interested reader is referred to Fung[5] for an in-depth discussion.

The acellular component is known as the extracellular matrix (ECM) and it provides the mechanical stability and much of the physical behaviour of interest. The ECM is a mesh of fibrous proteins, predominately collagen, along with a smaller proportion of elastin and even smaller proportions of fibronectin and laminin. The ECM mesh is permeated by a ground substance, a gelatinous mixture of water stabilised by macromolecules known as proteoglycans. These proteoglycans are highly hydrophilic allowing the retention of large amounts of water within tissues, contributing up to as much as 70–80% of the overall weight of cartilage and ~60–70% for other connective tissues.[4] It is this water retention in cartilage that confers great resistance to any compressive stress.[6]

Collagen and elastin

Collagen is the most abundant protein in the body and is the main structural component of all tissue, making up approximately 30% of the ECM.[6] In tissues, collagen forms a hierarchical structure. Single collagen molecules with a triple helical structure are created within

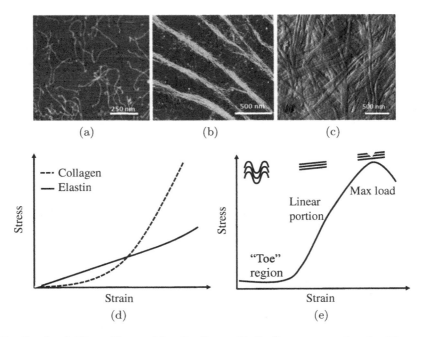

Fig. 5.   (a–c) The self-assembly of collagen fibrils (image reproduced with permission from Ref. 7); (d) comparison of the elastin and collagen stress–strain curves; (e) a typical stress–strain curve for tendon.

cells and secreted into the ECM (Fig. 5(a)). These helices self assemble into collagen fibrils (Fig. 5(b)). As they assemble, they do so in a staggered formation leading to a distinct banding which can be seen in fully formed fibrils (Fig. 5(c)). These fibrils are held together in tissues by strong covalent cross-links, such as disulphide bonds, and fibrils are further bundled into collagen fibres that exhibit a wavy or crimped structure (see Fung[5] for further details). The collagen molecule itself is relatively stiff, but the crimped configuration allows for a certain amount of extension of the tissue as a whole through fibre straightening. The packing of these fibres in tissues can confer anisotropy as well as other interesting properties. Parallel packing of fibres is seen in many load-transmitting tissues, such as tendon or ligament, and in the cornea, where layers of strictly parallel collagen fibres, oriented randomly to one another, confer optical transparency.

Elastin, by contrast, is the most elastic biosolid, and it forms a major component of blood vessels and lung tissue. Elastin is considered to operate as an entropic spring as, in its recoiled state, there is increased disorder of the surrounding water and of the polypeptide chain. Remarkably, elastin is able to extend to a stretch ratio of 1.6 whilst retaining an almost linear elastic stress–strain curve.[5]

The stress–strain curves of the collagen and elastin are shown in Fig. 5(d). The combination of the two fibre types gives rise to a typical stress–strain curve for soft tissue as depicted in Fig. 5(e). Imposing a mild stress on a tissue initially acts to straighten the collagen fibres with elastin storing a large amount of the strain energy; this is known as the toe region. The toe region is highly nonlinear and represents the normal physiological range for most tissues. Once the collagen fibres are aligned to the imposed stress and fully straightened a significant increase in tissue stiffness occurs, represented by the linear portion of the graph. Finally, the maximum stress indicates the state at which fibres break and permanent damage is caused to the tissue.

All tissues within the body vary in the relative amounts of fibrous proteins, ground substance and cellular component. This variation gives rise to the myriad of tissue properties present in the human body, from each tissue's structural stability, to its elasticity and compressive strength. There can be no single constitutive relation to describe all these behaviours and, instead, specific constitutive relations are formulated with a particular research question in mind. There are, however, a few assumptions that are widely adopted in most types of tissue models.

3.3. *Common assumptions*

Incompressibility

The assumption of incompressibility states that no volume change to the tissue occurs during a deformation. Alternatively, it can be thought of in terms of the density remaining constant, i.e., $\rho(\boldsymbol{x}, t) = \rho_0(\boldsymbol{X})$. Tissues are generally modelled as incompressible due to their high water content and because the packing distance of water molecules does not vary at biological temperature and pressure ranges. This assumption has a significant impact on the form of the constitutive equations and the range of parameter values. For instance, an incompressible material implies that Poisson's ratio approaches $\nu = 0.5$, Thus from Eqs. (3.2), λ must be much larger than μ and, indeed, both λ and K will tend to infinity in the incompressible limit.

In addition to these parameter constraints, the assumption of incompressibility also affects the form of the strain-energy density function (2.40). The local volume change from a deformation can be described in terms of the deformation gradient tensor, $F = F_{ij}$ in Eq. (2.5), by

$$\mathrm{d}x_1\mathrm{d}x_2\mathrm{d}x_3 = \det(F)\,\mathrm{d}X_1\mathrm{d}X_2\mathrm{d}X_3. \tag{3.3}$$

Hence, for an incompressible material $\det(\boldsymbol{F}) \equiv 1$ and, in terms of the three isotropic invariants of the right Cauchy–Green deformation tensor (2.38), equivalently $\mathcal{I}_3 \equiv 1$. This implies that a strain energy density for an incompressible material can only be expressed as a function of the first two invariants, i.e., $\mathcal{W}(\mathcal{I}_1, \mathcal{I}_2)$, in conjunction with $\mathcal{I}_3 \equiv 1$ as an additional constraint.

Isotropy

Isotropy is an assumption often made in the constitutive relations for rubbers and soft polymers. As tissues have a polymer-type structure many of these constitutive equations, and hence the isotropy assumption, are used in biomechanics. However, a crucial difference between rubber-type polymers and tissues is the level of organisation in the polymer chains or collagen fibres, respectively. As we have discussed, most tissues have a highly ordered arrangement of collagen fibres which is crucial to their function; on the other hand, most rubbers have a random arrangement of cross-linked polymer chains. This tissue anisotropy presents many difficulties for both theoretical and practical formulations of constitutive equations. Not only does anisotropy significantly increase the complexity of the constitutive relation (those constitutive relations from the fields of polymer science cannot simply be transferred across and applied), it also vastly increases the number of material parameters needed to describe the material. This complexity, in turn, means that determining these material properties in a reliable way becomes an intractable task in itself.

A compromise sometimes reached is to assume isotropy in order to create a tractable problem, but to then restrict the scale of investigation to scenarios where this assumption is considered to be valid. Sadly, this greatly reduces the predictive and descriptive power of such models. As more powerful large-scale computing becomes increasingly available, however, more and more constitutive relations that account for tissue anisotropy are being developed and implemented. A detailed discussion of anisotopy is outside the scope of this chapter but interested readers are referred to Sharpe.[3] We shall therefore continue this chapter with the assumption of isotropy that, at the very least, remains largely applicable to certain biological tissues such as cartilage and, importantly, to many tissue engineering materials.

3.4. *Hyperelasticity*

Hyperelasticity, introduced in Sec. 2.5, is the most common type of stress–strain relation used to describe tissue. This type of constitutive relation can account for large deformations and nonlinear elastic behaviour but notably it still assumes *strain-rate independence*. Hyperelastic relations were originally designed to describe the behaviour of polymers such as vulcanised rubber and other polymer foams, and have had much success in doing so.[8] Some common constitutive equations for polymers are given below:

$$\text{Neo-Hookean:} \quad \mathcal{W} = C_1(\mathcal{I}_1 - 3),$$
$$\text{Mooney–Rivlin:} \quad \mathcal{W} = C_1\,(\mathcal{I}_1 - 3) + C_2(\mathcal{I}_2 - 3),$$
$$\text{Gent:} \quad \mathcal{W} = -\frac{\mu J_m}{2}\ln\left(1 - \frac{\mathcal{I}_1 - 3}{J_m}\right).$$

Here, J_m in the Gent model represents some limiting finite value of molecular stretch allowing the model to mimic the *strain-stiffening* observed in biological tissues.[b]

---

[b]Obviously the relation is only valid for $\mathcal{I}_3 < J_m + 3$.

In addition to the problem of ordered, as opposed to random, fibre-orientation already mentioned when applying constitutive relations from polymer science to biological tissues, these formulations have a further drawback common to all hyperelastic models as explained in Sec. 2.5. Imposing a strain energy density function implies that deformations are a *reversible process* where energy stored in the stress material is regained on unloading. However, the straightening of the collagen fibres and the movement of the ground substance both dissipate energy; this phenomenon can be observed from the hysteresis in loading-unloading curves. Such an issue would have seriously impeded the application of hyperelastic models, were it not for the concept of *preconditioning*, introduced to biomechanics by Fung.[5] Preconditioning is the response of a tissue to cyclic loading within its physiological range. During cycling loading the hysteresis, which is initially large, becomes smaller as a steady state response is reached. Although there is never a complete absence of hysteresis, it becomes small enough that it can be neglected and a single elastic relation used to approximate the mean response. This observation led to the phrase *pseudo-elastic* being used to described such behaviour and Fung[5] developed a pseudo-elastic constitutive equation for pre-conditioned tissue based on this idea. Fung's pseudo-elastic strain energy function can be written as:

$$\mathcal{W} = \frac{1}{2}c[(e^Q - 1)], \qquad (3.4)$$

where in its most general form,

$$Q = C_{ijkl}\mathcal{E}_{ij}\mathcal{E}_{kl}. \qquad (3.5)$$

Note that both c and C_{ijkl} are material parameters that must be determined by fitting experimental data. This constitutive relation is able to predict a wide variation of observed tissue behaviours and has been extended in recent years to include compressibility and anisotropy.

3.5. *Constitutive relations based on microstructure*

The above formulations have all been based on experimental observations but, as noted at the beginning of this section, constructing a constitutive relation based on detailed knowledge of the tissue microstructure is an alternative approach. Lanir was the first to propose a strain energy density function where different fibre types, e.g., collagen and elastin, have individual strain energy densities and where the tissue as a whole is modelled based on the volume fraction Φ_k and an orientation density function $R_k(\phi, \theta)$ of each fibre type.[3] The strain energy of such a fibrous material can be expressed in the form

$$\mathcal{W} = \sum_{k=1,2} \Phi_k \int_0^\pi \int_0^\pi \mathcal{R}_k(\theta, \phi) w_k(\lambda) \sin\theta \, d\theta \, d\phi, \qquad (3.6)$$

where $w_k(\lambda)$ is the strain energy density for fibre type k in terms of the stretch ratio λ. The orientation density function is a structural property of the material such that $R_k(\theta, \phi) \sin\theta \, d\theta \, d\phi$ represents the proportion of fibres of type k with orientations in the range $(\theta, \theta + d\theta)$ and $(\phi, \phi + d\phi)$. Note that integration over all possible fibre orientations is only required over the hemisphere $0 \leq \theta, \phi \leq \pi$.

This general microstructural approach does have some benefits in practice, but it is difficult to implement owing to a lack of knowledge of the distribution functions. The approach, however, has been utilised in models of arterial layers and also validated against experimental data on fibre orientation in native bovine pericardium.

All of the constitutive relations described above are able to model certain aspects of tissue behaviour. However, strain-rate dependence is a common and important feature of tissue that cannot be correctly modelled using a hyper-elastic relation. To incorporate such behaviour, viscoelastic or poroelastic constitutive equations must be formulated, an introduction to these approaches will be covered in Secs. 5 and 6.

4.   Wave Propagation in Tissue

Sound is an unsteady mechanical perturbation that displaces particles in the medium in some oscillatory fashion. Audible sound is regarded as mechanical waves that propagate at frequencies ranging from 20 Hz to 20 kHz, representing the typical range of human hearing. Ultrasound is mechanical waves propagating at frequencies higher than the audible range and, for medical applications, this can typically extend to frequencies of the order of MHz. Both fluids and solids support compressive-type waves where the oscillations in density/displacement occur in the direction of travel (so-called longitudinal waves). Solids, unlike most fluids, also support wave propagation where the mechanical oscillations are perpendicular to the direction of travel; these are known as shear waves.

4.1. P-waves and S-waves

In this chapter, we shall assume that the mechanical waves generate only very small displacements (an assumption typical of acoustics) and, thus, the constitutive relation can be taken to be the general one for linear elasticity (2.17). Hence, we seek wave-like solutions to the unsteady Navier equation (2.18) where the displacement field takes the form

$$u_i = A_i \exp\left[\mathrm{i}\left(k_p X_p - \omega t\right)\right] \tag{4.1}$$

for constant vectors A_i and k_i (and where, of course, $k_p X_p = \boldsymbol{k} \cdot \boldsymbol{X}$) and with constant angular frequency ω. Substituting this wave-like form directly into (2.18) yields

$$\omega^2 \rho_0 A_i = (\lambda + \mu)\, k_i k_j A_j + \mu k_j k_j A_i. \tag{4.2}$$

In order to gain some proper insight from this equation, the vector A_i may be rewritten by expressing it uniquely in the form $A_i = ak_i + \epsilon_{ipq} b_p k_q$ for a given scalar a and a unique vector b_i that satisfies $b_i k_i = 0$. By then substituting this new expression for A_i into (4.2) one obtains

$$+\omega^2 \rho_0 \left(ak_i + \epsilon_{ipq} b_p k_q\right) = (\lambda + 2\mu)\, |\boldsymbol{k}|^2 ak_i + \mu |\boldsymbol{k}|^2 \epsilon_{ipq} b_p k_q; \tag{4.3}$$

two distinct types of wave can now be identified. The first type (present for $a \neq 0$) is known as a *P-wave*, which is a compressive type of wave as the mechanical oscillation is in the direction of the wave itself; this type of wave satisfies the relation $\rho_0 \omega^2 = (\lambda + 2\mu)\, |\boldsymbol{k}|^2$. The second type (present when $b_i \neq 0$) is known as an *S-wave*, where the oscillations in displacement are transverse to the direction of propagation, and this type satisfies the relation $\rho_0 \omega^2 = \mu |\boldsymbol{k}|^2$. The wave speeds, C_p and C_s, of the P-waves and S-waves respectively, given by $\omega/|\boldsymbol{k}|$, are therefore

$$C_p = \sqrt{\frac{\lambda + 2\mu}{\rho_0}} \quad \text{and} \quad C_s = \sqrt{\frac{\mu}{\rho_0}}. \tag{4.4}$$

These speeds, given the physical properties of materials in Sec. 3.1, must be both real and positive. Moreover, it is clear that $C_p > C_s$ in all solids. Indeed, this fact is well-known to seismologists who typically detect at least two distinct signals arriving one after another following a seismic disturbance, with the compressive P-wave arriving first followed by the S-wave.[c]

4.2. *Wave reflection and transmission*

Ultrasound imaging of soft tissue typically relies on the fact that waves are reflected at boundaries between different tissue types (skin, muscle, organs, bone). The return time of the reflected waves provides an indiction of the depth of any detected interfaces between different tissue types enabling the operator to build up a picture of the internal anatomy. Let's look at an example now of a plane P-wave, propagating from $x = -\infty$ towards $+\infty$ in an isotropic elastic medium with Lamé parameters λ_1, μ_1 and density ρ_1, that is incident on a flat interface at $x = 0$. This interface is the boundary with an adjoining elastic medium occupying the domain $x > 0$ with different Lamé parameters λ_2, μ_2 and density ρ_2. If one were to naïvely attempt to solve this problem in a similar way to a plane acoustic wave, then we would assume that there would be a P-wave reflected by the interface with relative amplitude R_p and a transmitted refractive P-wave into the medium occupying $x > 0$ with relative amplitude T_p. Thus, our displacement field, $\boldsymbol{u}(x, y) = (u, v)^T$, in the

---

[c]The S-wave signal is subsequently followed by other signals from surface waves (mentioned below) that travel more slowly.

domain would take the form

$$
\boldsymbol{u}(\boldsymbol{x}) =
\begin{cases}
\begin{bmatrix} u_1(x,y) \\ v_1(x,y) \end{bmatrix} e^{-i\omega t} & \text{for } x < 0, \\[2em]
\begin{bmatrix} u_2(x,y) \\ v_2(x,y) \end{bmatrix} e^{-i\omega t} & \text{for } x > 0,
\end{cases}
\tag{4.5}
$$

where, given an angle of incidence α and P-wave speeds C_{p1} and C_{p2} in $x < 0$ and $x > 0$ respectively,

$$
\begin{bmatrix} u_1 \\ v_1 \end{bmatrix} = \begin{bmatrix} \cos\alpha \\ \sin\alpha \end{bmatrix} e^{i\frac{\omega}{C_{p1}}(x\cos\alpha + y\sin\alpha)} + R_p \begin{bmatrix} -\cos\beta \\ \sin\beta \end{bmatrix} e^{i\frac{\omega}{C_{p1}}(-x\cos\beta + y\sin\beta)},
\tag{4.6}
$$

$$
\begin{bmatrix} u_2 \\ v_2 \end{bmatrix} = T_p \begin{bmatrix} \cos\gamma \\ \sin\gamma \end{bmatrix} e^{i\frac{\omega}{C_{p2}}(x\cos\gamma + y\sin\gamma)}.
\tag{4.7}
$$

The necessary matching conditions at the interface are both continuity of the displacement, $u_1(0, y) = u_2(0, y)$, $v_1(0, y) = v_2(0, y)$, and continuity of the applied stress, given explicitly here as

$$
2\mu_1 \frac{\partial u_1}{\partial x} + \lambda_1 \left(\frac{\partial u_1}{\partial x} + \frac{\partial v_1}{\partial y} \right) = 2\mu_2 \frac{\partial u_2}{\partial x} + \lambda_2 \left(\frac{\partial u_2}{\partial x} + \frac{\partial v_2}{\partial y} \right) \quad \text{at } x = 0,
\tag{4.8}
$$

$$
\mu_1 \left(\frac{\partial u_1}{\partial y} + \frac{\partial v_1}{\partial x} \right) = \mu_2 \left(\frac{\partial u_2}{\partial y} + \frac{\partial v_2}{\partial x} \right) \quad \text{at } x = 0.
\tag{4.9}
$$

Matching the displacement fields for all y at $x = 0$ determines uniquely the respective angles of reflected and transmitted waves via Snell's law, thus

$$
\frac{\sin\alpha}{C_{p1}} = \frac{\sin\beta}{C_{p1}} = \frac{\sin\gamma}{C_{p2}}.
\tag{4.10}
$$

However, four equations remain for the amplitudes R_p and T_p, meaning that the problem is overspecified unless we have an incoming wave at normal incidence (where $\sin\alpha = 0$). Thus, our current form of solutions (4.6) and (4.7) are only valid for $\alpha = 0$, in which case there are just two, rather than four, non-trivial equations yielding the solution

$$
R_p = \frac{Z_2 - Z_1}{Z_1 + Z_2}, \quad T_p = \frac{2Z_1}{Z_1 + Z_2}, \quad \text{for } Z_1 = \rho_1 C_{p1} \text{ and } Z_2 = \rho_2 C_{p2}.
\tag{4.11}
$$

Z_1 and Z_2 are well known in classical acoustics as the *characteristic impedances* of each medium.

For an incident wave at an oblique angle ($\sin \alpha \neq 0$), the only way the problem can remain well-posed is to allow *mode-conversion* by adding a reflected shear wave of the form

$$+R_s \begin{bmatrix} \sin \delta \\ \cos \delta \end{bmatrix} e^{i \frac{\omega}{c_{s1}} (-x \cos \delta + y \sin \delta)} \qquad (4.12)$$

with unknown amplitude R_s and angle δ to (4.6) for $x < 0$, and a refracted shear wave of the form

$$+T_s \begin{bmatrix} -\sin \varepsilon \\ \cos \varepsilon \end{bmatrix} e^{i \frac{\omega}{c_{s2}} (x \cos \varepsilon + y \sin \varepsilon)} \qquad (4.13)$$

with unknown amplitude T_s and angle ε to (4.7) for $x > 0$. Once again, given the two S-wave speeds C_{s1} and C_{s2} in $x < 0$ and $x > 0$ respectively, Snell's law then specifies all the angles, including $\alpha = \beta$ and

$$\frac{\sin \alpha}{C_{p1}} = \frac{\sin \gamma}{C_{p2}} = \frac{\sin \delta}{C_{s1}} = \frac{\sin \varepsilon}{C_{s2}}. \qquad (4.14)$$

The four equations matching displacement and stress at the interface that determine the unknown amplitudes of reflected and transmitted waves (R_p, R_s, T_p, T_s) are known as the Knott-Zoeppritz equations [9, p140] and have been extensively studied in seismology. These equations can be put into matrix form but, unfortunately, in this form they do not provide much intuitive insight. As a consequence, various approximations for mode-conversion reflection and transmission have been derived over the years.

Note that as the S-wave speeds are always lower than P-wave speeds one can expect from (4.14) the angle of each S-wave to the normal, δ and ε, to be smaller than the angle of the corresponding P-wave, α and γ, in each medium as shown in Fig. 6. While four reflected and refracted waves are typically expected for oblique angles of incidence, in the case of $C_{p2} > C_{p1}$ this is only true for $0 < \alpha < \sin^{-1}(C_{p1}/C_{p2})$ because, once this critical angle is reached, a P-wave is no longer transmitted into $x > 0$. Instead, an evanescent so-called head wave is generated that travels along the interface $x = 0$. Indeed, if the S-wave speed in $x > 0$ also happens to be faster than the incident

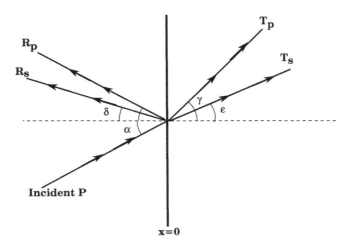

Fig. 6.   An incident P-wave impinging on a flat interface between two elastic media.

wave speed, i.e., $C_{s2} > C_{p1}$, then a second critical angle also occurs at $\alpha = \sin^{-1}(C_{p1}/C_{s2})$ beyond which total internal reflection occurs.

Other important waves in solid mechanics are surface waves that are typically dispersive, decay exponentially with depth, and have lower propagation speeds than the P-waves and S-waves which propagate through the interior of the solid. Two well-known types of surface waves are Rayleigh waves, which generate a ground roll, and Love waves, which generate horizontal surface shifts.[10] Surface waves tend to be the primary cause of the significant movements people experience during earthquakes.

4.3.   *Imaging techniques*

In ultrasound, despite the mode conversion observed in the above example, it is common to ignore S-waves altogether and model soft tissue practically as a fluid supporting P-wave propagation only. This is because (i) the equations are easier, (ii) ultrasound transducers tend to emit compressive P-waves only and (iii) because S-waves are more strongly attenuated by soft tissue and thus do not have the same degree of tissue penetration. Additionally, the dramatically differing sound speeds mean that the reflected P-wave always arrives back at the detector well before any signal from the strongly attenuated S-waves, thus allowing them to be easily distinguished.

Attenutation of ultrasound waves arises from a myriad of absorption and scattering phenomena. Absorption mechanisms include viscous, chemical and thermal effects. Scattering phenomena, on the other hand, arises from small scale inhomogeneities, of size typically of the order of the ultrasound wavelength or smaller, embedded in the tissue. The overall attenuation of ultrasound is usually modelled by including an exponentially-decaying term to the mechanical wave of the form $e^{-\beta x}$ where x is the distance travelled and β is known as the *absorption coefficient*. The absorption coefficient is frequency dependent and has dimensions of the reciprocal of the length (e.g., cm^{-1} or m^{-1}); it is often expressed in terms of decibels per centimetres via the conversion

$$20 \log_{10} \left(\frac{A_0\, e^{-\beta}}{A_0} \right) \approx -8.686\beta \text{ dBcm}^{-1}. \tag{4.15}$$

Ultrasound attenuation can be incorporated into a numerical simulation by adopting a complex wavenumber where $\text{Im}(k) = \pm\beta$ (with the sign depending on the sign of the $i\omega t$ exponent). Experiments have shown that attenuation of ultrasound in biological tissues varies approximately linearly with frequency[11] over a range 1–50 MHz. Such behaviour contrasts markedly with the typical attenuation of sound observed in viscous fluids, which tends to vary as the square of the frequency.

The success of ultrasound in medical imaging compared to other modalities (aside from the fact that it is non-ionising) is because compressive sound speeds across the most common soft tissue types (fat, muscle, organs, brain) remain broadly similar while the characteristic impedances ρZ are distinctly different[12]; such properties generate significant reflections at interfaces while enabling transit times to be straightforwardly determined for calculating interface depth. These simplifying assumptions, however, are not valid when imaging bone due to the significantly disparity in wave speed, the greater amount of mode conversion, the lower S-wave attenuation and the higher scattering caused by bone heterogeneity (particularly due to its porous nature) (see Ref. 13, pp. 41–52).

While traditional ultrasound imaging techniques neglect S-waves, magnetic resonance elastography and sonoelastography techniques generate and utilise shear waves to measure the elastic properties of soft tissue. In particular, shearwave dispersion ultrasound vibrometry (SDUV) is a recently developed ultrasound technique using amplitude-modulated ultrasound that measures shear wave dispersion to characterise important properties of human tissue.[14] The shear modulus and viscoelastic properties of muscle, arterial vessels and organs, such as the liver, can be determined by SDUV with excellent spatial and temporal resolution, thus potentially enabling this technique to spatially differentiate between areas of normal and pathological tissue behaviour.

5. Viscoelasticity

All tissues within the body exhibit some degree of time-dependent stress–strain behaviour, in addition to the elastic behaviour we have already discussed. The viscous time-dependent response of soft tissue can be important for the tissue's proper function and changes in its viscoelastic properties can be an indicator for disease such as short sightedness, liver fibrosis and cancer.

The molecular mechanisms of the viscous properties in tissue are highly complex and, even now, are not fully understood. It has been shown that collagen fibrils do not appear to exhibit any *intrinsic* viscoelasticity, but that stress relaxation of the whole tissue could result from a slipping mechanism between collagen fibrils and proteoglycans. Simulating the viscoelastic properties of tissue adds a significant amount of complexity to a numerical model and, therefore, it should first be thoroughly considered whether accurately capturing such behaviour in a model is necessary for the situation of interest. When making this decision, it is crucial to consider the characteristic time of the viscoelastic response relative to the timescale of the experiment in question; the ratio of these timescales is a non-dimensional parameter known as the Deborah number.[10] For instance, if an experiment happens over a timescale vastly different from the characteristic viscoelastic response time then a model such as the hyperelastic or pseudoelastic models from the previous sections may be adequate and, indeed, less cumbersome.

5.1. *Types of viscoelastic behaviour*

Three types of viscoelastic behaviour are used to quantify parameters and fit an appropriate constitutive model:

- *Creep testing* — the application of an instantaneous and constant stress. For certain viscoelastic materials, if the strain is measured over an appropriate time period it will be seen to monotonically increase over time.
- *Stress relaxation* — the application of an instantaneous and constant strain. For certain viscoelastic materials a decrease in the stress over time will be observed.
- *Dynamic loading* — the application of a sinusoidally-varying stress or strain. For a viscoelastic material, there will typically be a lag between the stress σ and strain ε with an associated phase angle δ, thus

$$\varepsilon = \varepsilon_0 \cos(\omega t),$$
$$\sigma = \sigma_0 \cos(\omega t + \delta).$$

Using the standard techniques of harmonic oscillators we can write the stress in complex notation as

$$\sigma^* = \left(\sigma_0' + i\sigma_0''\right) e^{+i\omega t}. \tag{5.1}$$

Two moduli are associate with Eq. (5.1): $G' = \sigma_0'/\varepsilon_0$ which is known as the real or storage modulus, and $G'' = \sigma_0''/\varepsilon_0$ which is the loss or imaginary modulus. In a purely elastic material the storage modulus will be equal to the shear modulus μ. Experimental testing of various tissues presented by Fung[5] shows the variation in the viscoelastic response of soft tissues. There are a variety of constitutive models which can be used to fit this type of data, and the two types we focus on here are the mechanistic and quasilinear viscoelastic models.

5.2. *Mechanistic models*

Mechanistic models provide mechanical analogs for the different components of the tissue's response: the elastic components of the

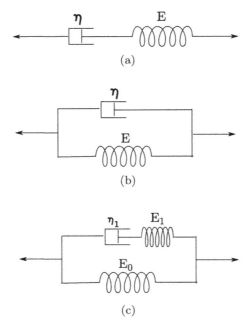

Fig. 7. Mechanistic models: (a) Maxwell model; (b) Voigt model; (c) Kelvin or Standard linear model

material are represented by a series of springs and the viscous components by dashpots as shown in Fig. 7. In all of the three models shown in Fig. 7, the springs are linearly elastic with spring constant E and the dashpots are linear with viscous coefficient η. In each case the stress–strain relation of each element is given by:

$$\sigma_e = E\varepsilon \quad \text{and} \quad \sigma_v = \eta\frac{\mathrm{d}\varepsilon}{\mathrm{d}t}. \tag{5.2}$$

In the Maxwell model, shown in Fig. 7(a), the same tension is transmitted to both the spring and the dashpot with the displacement being the sum of the elastic and viscous components. Hence,

$$\varepsilon = \varepsilon_e + \varepsilon_v, \quad \sigma = E\varepsilon_e = \eta\frac{d\varepsilon_v}{dt} \tag{5.3}$$

$$\text{and} \quad \frac{\mathrm{d}\varepsilon}{\mathrm{d}t} = \frac{1}{E}\frac{\mathrm{d}\sigma}{\mathrm{d}t} + \frac{\sigma}{\eta}. \tag{5.4}$$

Rewriting Eq. (5.4), using $\tau = \eta/E$, we can define the constitutive equation

$$E\frac{d\varepsilon}{dt} = \frac{d\sigma}{dt} + \frac{1}{\tau}\sigma. \tag{5.5}$$

In stress relaxation, an instantaneous strain ε_0 is applied at $t = 0$ and kept constant. The time-dependent stress response is found by integrating Eq. (5.5) with $d\varepsilon/dt = 0$ yielding

$$\sigma(t) = E\varepsilon_0 e^{-\frac{t}{\tau}}. \tag{5.6}$$

From this it is easy to see that τ refers to characteristic time known as the *relaxation time*. If we divide the stress by the initial strain, we get the so-called *relaxation modulus*,

$$G(t) = Ee^{-\frac{t}{\tau}}. \tag{5.7}$$

$G(t)$ is a characteristic function that describes how the stress relaxes for a given material.

The same process can be repeated for the application of an instantaneously applied stress. In this case, we find:

$$J(t) = \left(\frac{t}{E} + \frac{1}{\eta}\right), \tag{5.8}$$

where $J(t)$ is called the *creep function*. These two functions can be derived for all three mechanistic models in Fig. 7 as above. These three models are particular examples of linear systems and a generalised formulation for all spring-dashpots models can be developed using the Boltzmann superposition principle, leading to the integral forms[10]

$$\varepsilon(t) = \int_{-\infty}^{t} J(t - t')\frac{d\sigma(t')}{dt'}dt', \tag{5.9}$$

$$\sigma(t) = \int_{-\infty}^{t} G(t - t')\frac{d\varepsilon(t')}{dt}dt'. \tag{5.10}$$

Generalisation of the relaxation and creep functions to include multiple relaxation times is easily done via a summation of single exponential functions

$$G(t) = \sum_{i=1}^{n} = G_i e^{-\frac{t}{\tau}}. \tag{5.11}$$

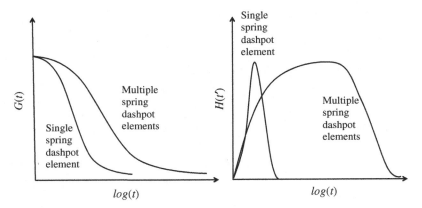

Fig. 8.   A graphical representation of the relaxation function $G(t)$ and relaxation spectrum function $H(t)$.

Moreover, one further refinement of the above models is to extend the relaxation function (5.11) to be a continuous function $H(\tau)$, known as the continuous relaxation spectrum (Fig. 8), given by

$$G(t) = \int_{0}^{\infty} \frac{H(\tau)}{\tau} e^{-\frac{t}{\tau}} d\tau. \tag{5.12}$$

Finally, we note that these generalised formulations are equally valid for 3D materials where, in tensor form, we find:

$$\varepsilon_{ij}(\boldsymbol{x}, t) = \int_{-\infty}^{t} J_{ijkl}(\boldsymbol{x}, t - t') \frac{\partial \sigma_{kl}(\boldsymbol{x}, t')}{\partial t'} dt', \tag{5.13}$$

$$\sigma_{ij}(\boldsymbol{x}, t) = \int_{-\infty}^{t} G_{ijkl}(\boldsymbol{x}, t - t') \frac{\partial \varepsilon_{kl}(\boldsymbol{x}, t')}{\partial t'} dt'. \tag{5.14}$$

5.3.   *Quasilinear viscoelasticity*

While the above models are useful, they are **linear** viscoelastic models and are, therefore, only valid for small deformations. For tissues where large finite deformations occur the elastic response is inherently nonlinear. For this reason, the most commonly adopted viscoelastic model in tissue mechanics is Fung's quasilinear viscoelasticity (QLV).[5] Fung proposed that a relaxation modulus could be separated into time-dependent and strain-dependent parts as follows,

$$G(\varepsilon, t) = g(t)\frac{\partial \sigma^e(\varepsilon)}{\partial \varepsilon}. \tag{5.15}$$

The function $g(t)$ is known as the reduced relaxation function. $\sigma^e(\varepsilon)$, on the other hand, represents the *instantaneous* elastic response to a change in strain. QLV retains linearity in the time response but allows for a nonlinear elastic response through a nonlinear constitutive relation $\sigma^e(\varepsilon)$; relations of the form $\sigma^e(\varepsilon) = A\left(e^{B\varepsilon} - 1\right)$ are typically used. The actual stress of the viscoelastic material at a given time, $\sigma(t)$, can then be obtained by assuming superposition of all previous constituents of the material's strain history, thus

$$\sigma(t) = \int_{-\infty}^{t} g(t - t')\frac{\partial \sigma^e(\varepsilon)}{\partial \varepsilon}\frac{\partial \varepsilon}{\partial t'}dt'. \tag{5.16}$$

Many forms of the reduced relaxation function are possible and, although the theory is and has been widely used, recent experimental results show it is unable to adequately model some of the more complex behaviours observed, including the different short and long time-scale viscosities that arise from different molecular mechanisms.[3] There have been other numerous attempts to model nonlinear viscoelastic properties and those interested are referred to two review papers: one by Verdier[15] and one by Cowin and Doty.[6]

6. Poroelasticity

6.1. *Historical outline*

The last tissue model covered in this chapter is poroelasticity. In the previous section, we considered a viscoelastic model of tissue to account for the time-dependent stress–strain responses. While viscoelastic models have been successful in simulating certain biological scenarios, they do not provide much insight into the biological mechanisms underlying the time-dependent behaviours. Poroelasticity seeks to model these time-dependent responses in a way that accords with our current understanding of tissue microstructure.[3] It achieves this by modelling the behaviours of a solid porous matrix and a pore fluid as separate phases, which are coupled through the transfer of mass and momentum. We already know tissues consist of a matrix of fibrillar proteins pervaded by a water-rich ground substance. With such an approach, both the elastic and time-dependent behaviours of a tissue can be modelled by the combination of a suitable elastic model for the fibrillar proteins and the fluid dynamic response of the ground substance. This approach to tissue modelling was first transferred from geology to biomechanics in the 1970s and 1980s and, subsequently, the work expanded into so-called biphasic theory.

There are broadly two formulations of poroelasticity. One is a solid-mechanics type formulation that is based on Biot's original work published in 1941, which has been advanced by a number of key contributors including Rice, Cleary and Rudnicki. The second is mixture or biphasic theory that stems from a fluid mechanics standpoint and is formulated in terms of diffusion equations; this approach can be attributed to Fick and Stefan.[6] The Biot formulation describes the poroelasticity in terms of a representative volume element where homogenisation procedures are used to derive a valid continuum model from the volume element. In order for this homogenisation approach to be valid it is implied that the dimension of the volume element must be much larger than the pore size. Mixture theory is formulated based on the weighted average fluxes of the mixture components passing through a fixed spatial point.

An in-depth discussion of the two formulations can be found in Ref. 6. The two formulations have been shown to be equivalent, and there is increasing interest in the expansion of mixture theory to include more than two phases.[5, 6]

6.2. *Fundamental concepts*

In a poroelastic material, both the fluid-filled pores and the solid material are fully interconnected. The elastic component is characterised by a set of solid-mechanics parameters and the pore fluid

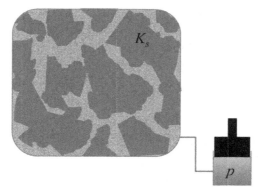

Fig. 9. A schematic of a poroelastic material comprising an interconnected pore space occupied by fluid that is surrounded by a solid elastic solid matrix or skeleton.

characterised by a pore pressure denoted by p (Fig. 9). The derivation covered here follows that of Ref. 16 which, in turn, is based on Biot's original formulation. For the interested reader, Ref. 17 offers a consistent but slightly different approach. The derivation begins by introducing the basic physical concepts, assumptions and parameters before deriving the governing equations and using linear elastic constitutive equations to close the problem. Nonlinear poroelastic constitutive relations used in tissue mechanics are briefly described at the end of the section.

The porosity is defined as $\Phi = \frac{V_p}{V}$, where V is the entire volume of the poroelastic material and V_p is the volume of the interconnected pore space (thus $0 < \Phi < 1$). The variation of fluid content ζ is the change in fluid volume per unit volume of solid matrix. Thus $\zeta > 0$ represents a net gain of fluid by the poroelastic material, whereas $\zeta < 0$ represents a net loss (compared to some reference state).

The so-called *drained* and *undrained* responses are two types of limiting conditions that are crucial to the understanding and characterisation of poroelastic behaviour. In drained conditions, the interstitial fluid can flow freely into and out of the solid matrix and so the pore pressure must be zero. This means that only the elastic portion of the material need be considered and this is equivalent to the long-time scale behaviour of the material. The undrained response describes the situation where fluid cannot escape the surrounding elastic matrix and so $\zeta \equiv 0$. As a consequence, any external pressure change causes an immediate response in the pore pressure and, therefore, this represents the instantaneous or short-time-scale behaviour. The undrained condition gives rise to a set of parameters, analogous to the elastic parameters but with the subscript u (for instance, ν_u is the undrained poisson ratio and K_u is the undrained bulk modulus). The undrained parameters can be related to their drained or the purely elastic counterparts via the Biot-Willis coefficient α and the porosity Φ as described below.

Parameters

Notation in poroelasticity is not consistent within the available literature as different parameters are chosen and adopted by different authors. Hence, there are no agreed, well-defined symbols associated with the coefficients. For the parameters used in this section, we have tried to be consistent with the notation and formulation found in Ref. 16 where a full list can be found.

> R: the *unconstrained specific storage coefficient*, which specifies the variation of fluid content in a control volume due to an applied change in the pore pressure p.

> $1/H$: *the poroelastic expansion coefficient*, which describes the change in bulk volume due to pore pressure whilst holding the stress constant.

> α: *the Biot-Willis coefficient* (mentioned above) is defined by $\alpha = K/H$. It describes the ratio of the fluid volume gained or lost due to a bulk volume change.

> B: the *Skepton coefficient*, which is defined by $B = R/H$, is the ratio of pore pressure to total pressure in the undrained condition.

> M: the *Biot modulus*, is given by $M = \frac{K_u - K}{\alpha^2}$.

> κ: the permeability of the matrix, which is used in Darcy's law for fluid transport. It is defined as k/η where k is the intrinsic permeability and η is the viscosity of the fluid.

Effective stress

Pore fluid moving through the elastic matrix generates shear and normal stresses that would need to be analysed by a detailed fluid dynamics approach, making the poroelastic formulation vastly complicated. *Terzaghi's principle* states that on a scale of many pores the stress in the fluid can be averaged to an isotropic pore pressure p. The total stress in the material is then given by:

$$\sigma_{ij} = -p\delta_{ij} + \sigma_{ij}^e, \tag{6.1}$$

where σ_{ij}^e is the elastic matrix contribution. This principle is of critical importance to obtaining a tractable model and is used in all poroelastic formulations.

6.3. *Governing equations*

In order to derive a model for the stress–strain behaviour of a poroelastic tissue, three types of governing laws are needed: (i) conservation laws of mass and momentum for the solid and the fluid; (ii) transport laws for the fluid through the solid; and (iii) constitutive relations for both the solid and the fluid. The solid porous component is described as a linear elastic material based on the standard notation of a displacement $u(X, t)$ as introduced in Sec. 2. The fluid can be described in terms of a *specific discharge* vector q_i which is defined as the local rate of fluid volume crossing a unit area of porous solid.

(i) *Conservation laws*

Conservation of momentum for the porous elastic material is given by

$$\frac{\partial \sigma_{ij}}{\partial X_j} = -F_i, \tag{6.2}$$

where F_i is the body force per unit volume of the *bulk material*. The continuity equation derived from conservation of mass for the fluid can be written as

$$\frac{\partial \zeta}{\partial t} = -\frac{\partial q_i}{\partial X_i} + \gamma, \tag{6.3}$$

where γ, if required, represents a source density (rate of injected fluid per unit volume).

(ii) *Transport law*

Darcy's law governs the flow of fluid through a porous solid. It can be formulated as

$$q_i = -\kappa \left(\frac{\partial p}{\partial X_i} - f_i \right), \tag{6.4}$$

where f_i is the body force per unit volume of the *fluid*. Substitution of Eq. (6.3) into Darcy's law yields

$$\frac{\partial \zeta}{\partial t} = \kappa \nabla^2 p + \gamma - \kappa \frac{\partial f_i}{\partial X_i}. \tag{6.5}$$

(iii) *Constitutive relations*

Constitutive relations for the solid and fluid components can be combined with (6.5) above to produce a Navier-type equation coupled to a diffusion equation. If the constitutive relation for linear elasticity is taken for the solid component, so that the linear strain tensor is given by

$$\varepsilon_{ij} = \frac{1}{2} \left(\frac{\partial u_i}{\partial X_j} + \frac{\partial u_j}{\partial X_i} \right), \tag{6.6}$$

then the constitutive equations for a linear poroelastic material in their simplest form can be written as

$$\varepsilon_{ij} = \frac{\sigma_{ij}}{2\mu} - \left(\frac{1}{6\mu} - \frac{1}{9K} \right) \delta_{ij}\sigma_{kk} + \frac{1}{3H}\delta_{ij}p, \tag{6.7}$$

$$\zeta = \frac{\sigma_{kk}}{3H} + \frac{p}{R}; \tag{6.8}$$

we remark here that ζ thus represents in some way the strain of the interstitial fluid as opposed to ε_{ij} which represents the strain of the solid matrix.

Equations (6.7) and (6.8) can be rewritten in terms of the total stress σ_{ij} and pore pressure p:

$$\sigma_{ij} = 2\mu\varepsilon_{ij} + \left(K - \frac{2\mu}{3} \right) \delta_{ij}\varepsilon_{kk} - \alpha\delta_{ij}p, \tag{6.9}$$

$$p = M(\zeta - \alpha\varepsilon_{kk}), \tag{6.10}$$

where the volumetric strain is $\varepsilon_{kk} = (\sigma_{kk}/3 + \alpha p)/K$. Combining (6.5) with (6.10) yields a diffusion type equation, which is equivalent to the storage equation of Ref. 17, given by

$$\frac{\partial p}{\partial t} - \kappa M \, \nabla^2 p = -\alpha M \frac{\partial \varepsilon}{\partial t} + M \left(\gamma - \frac{\partial f_i}{\partial X_i} \right). \tag{6.11}$$

Finally, combining (6.6), (6.7) and (6.2) leads to a Navier-style equation of the form

$$\mu \, \nabla^2 u_i + \frac{\mu}{(1 - 2\nu)} \frac{\partial^2 u_k}{\partial X_k \partial X_i} = \alpha \frac{\partial p}{\partial X_i} - F_i. \tag{6.12}$$

Nonlinear poroelasticity

An improvement on the above formulation for tissue modelling can be made by replacing the linear elastic constitutive relation with a hyper-elastic strain-energy density (Sec. 3.4). The increment of work $\delta \mathcal{W}$ in a poroelastic material is the sum of the elastic and fluid components and, from linear theory, this implies

$$\delta \mathcal{W} = \sigma_{ij} \mathrm{d}\varepsilon_{ij} + p \mathrm{d}\zeta; \tag{6.13}$$

therefore $\frac{\partial \mathcal{W}}{\partial \varepsilon_{ij}} = \sigma_{ij}$ and $\frac{\partial \mathcal{W}}{\partial \zeta} = p$. However, for finite rather than infinitesimal deformations the Lagrangian formulation, as detailed in Sec. 2.4, should now be adopted with the Cauchy stress replaced by the second Piola–Kirchhoff stress tensor. Thus, our equation for total stress (6.1) becomes

$$\mathcal{S}_{ij} = -p \det(F) \mathcal{C}_{ij}^{-1} + \mathcal{S}_{ij}^e, \tag{6.14}$$

where $\mathcal{C}_{ij}^{-1}$ is the inverse of the right Cauchy–Green deformation tensor as defined in Sec. 2.1. The Lagrangian stress component for the elastic matrix can be derived from an effective strain energy density function $\mathcal{W}^e$ by the relation

$$S_{ij}^e = \frac{\partial \mathcal{W}^e}{\partial \mathcal{E}_{ij}} \tag{6.15}$$

and several forms of effective strain energy density function can be used, one example being

$$\mathcal{W}^e = \frac{1}{2} C_0 (e^{C_{ijkl} \mathcal{E}_{ij} \mathcal{E}_{kl}} - 1). \tag{6.16}$$

In addition to the modifications to the constitutive relation, for large deformation porohyperelasticity, the permeability of the matrix (which appears in Darcy's law) is now assumed to be a function of the deformation, thus $\kappa(\det(F))$. One suggested form of this relation is

$$\kappa = \kappa_0\, e^{M(\det(F)-1)}. \tag{6.17}$$

For those interested, a recent in-depth analysis of this and other porosity functions applied to a number of hyperelastic relations is given by Ref. 18.

7. Exercises

(1) Show that the strain tensor $\mathcal{E}_{ij}$ is identically zero for any rigid body displacement of the form $x_i = c_i + Q_{ij}X_j$, where c_i is a constant vector and Q_{ij} is an orthogonal matrix. However, show that the linearised strain tensor ε_{ij} is *not* identically zero under such displacement. Determine for what form of displacement $\boldsymbol{u}$ the linearised strain is identically zero and explain the discrepancy.

(2) Show that a constitutive relation of the form $\mathcal{S}_{ij} = (\mathcal{I}_3 - 1)\delta_{ij}$ leads to energy generation under a periodic loading and unloading cycle. Hence, confirm that such a relation cannot satisfy the conditions of hyperelasticity (see Ref. 10, Sec. 5.3.4 for an example).

(3) For the simple shear $x_1 = X_1 + \alpha X_2$, $x_2 = X_2$ and $x_3 = X_3$ (for a given value of α), find the deformation gradient F_{ij} and the right Cauchy–Green deformation tensor $\mathcal{C}_{ij}$. Show that the eigenvalues of $\mathcal{C}_{ij}$, λ_1, λ_2 and λ_3, satisfy

$$\lambda_1 + \lambda_2 = 2 + \alpha^2, \quad \lambda_1\lambda_2 = 1 \quad \text{and} \quad \lambda_3 = 1.$$

(4) A spherical cavity within a tissue phantom of initial radius a and at pressure p_0 expands to a larger radius b due to an increase in internal pressure. The expansion is in the radial direction only, and the tissue phantom is incompressible. Using Cauchy's equation in spherical coordinates and the Gent hyperelastic constitutive equation with $J_{lim} = \infty$, show that the pressure in the

cavity is given by

$$P = P_0 + \frac{\mu}{2} \left[5 - 4 \left(\frac{b}{a}\right)^{-1} - \left(\frac{b}{a}\right)^{-4} \right].$$

Hint: see Zhu et al. (2011) Journal of Adhesion 87(5), pp. 466–481.

(5) Suppose a plane P-wave of known amplitude is incident on a flat rigid boundary (i.e., on which the displacement is zero). Determine the reflected wave field and hence show that mode conversion occurs for oblique angles of incidence. Repeat for an incident S-wave.

(6) Suppose we have a viscoelastic tissue that has an instantanous elastic response given by

$$\sigma^e(\varepsilon) = A(e^{B\varepsilon} - 1),$$

where A and B are given constants and ε is the strain. A linearly increasing strain is applied to the tissue such that, at any time after $t = 0$, the strain is given by $\varepsilon(t) = \gamma t$ for some $\gamma > 0$. Given the reduced relaxation function $g(t) = \sum_{i=1}^{N} h_i e^{-\alpha_i t}$, find the analytical expression for the stress $\sigma(t)$ at some time $t > 0$ using the QLV formulation. How does the tissue response differ if for time $t > t_0 > 0$ the strain stops increasing and stays at a constant level, $\varepsilon(t > t_0) = \gamma t_0$?

(7) Derive both the storage equation (6.11) and Navier equation (6.12) for a linear poroelastic material. Then derive a diffusion-type equation in terms of ζ and show that the diffusion coefficient $\mathcal{D}_\zeta$ is given by

$$\mathcal{D}_\zeta = \frac{2\kappa\mu(1-\nu)(\nu_u-\nu)}{\alpha^2(1-2\nu)^2(1-\nu_u)}.$$

8. Solutions to Selected Problems

(3) First, we calculate the deformation gradient F_{ij} and then obtain the right Cauchy–Green deformation tensor:

$$C_{ij} = F_{ki}F_{kj} = \begin{pmatrix} 1 & \alpha & 0 \\ \alpha & (1+\alpha^2) & 0 \\ 0 & 0 & 1 \end{pmatrix}.$$

To find the eigenvalues we solve $\det(\lambda\delta_{ij} - \mathcal{C}_{ij}) = 0$, which yields the characteristic polynomial

$$(\lambda - 1)\left(\lambda^2 - \lambda(2 + \alpha^2) + 1\right) = 0.$$

The eigenvalues are therefore

$$\lambda_{1,2} = 1 + \frac{\alpha^2}{2} \pm \sqrt{\left(1 + \frac{\alpha^2}{2}\right)^2 - 1} \quad \text{and} \quad \lambda_3 = 1,$$

which satisfy $\lambda_1 + \lambda_2 = 2 + \alpha^2$ and $\lambda_1 \lambda_2 = 1$.

(5) We assume that the displacement field in the medium occupying $x < 0$ is given by the sum of (4.6) and (4.12) as in Sec. 4:

$$\begin{bmatrix} u_1 \\ v_1 \end{bmatrix} = \begin{bmatrix} \cos\alpha \\ \sin\alpha \end{bmatrix} e^{i\frac{\omega}{C_{p1}}(x\cos\alpha + y\sin\alpha)}$$

$$+ R_p \begin{bmatrix} -\cos\beta \\ \sin\beta \end{bmatrix} e^{i\frac{\omega}{C_{p1}}(-x\cos\beta + y\sin\beta)},$$

$$+ R_s \begin{bmatrix} \sin\delta \\ \cos\delta \end{bmatrix} e^{i\frac{\omega}{C_{s1}}(-x\cos\delta + y\sin\delta)},$$

where C_{p1} and C_{s1} are the P-wave and S-wave speeds respectively. Setting $u_1 = 0$ and $v_1 = 0$ for $x = 0$ leads to $\alpha = \beta$ and $\sin\delta = \frac{C_{s1}}{C_{p1}}\sin\alpha$ and thus $\delta < \alpha$ as $C_{p1} > C_{s1}$. This leaves two equations for the unknown amplitudes of the reflected P- and S-waves:

$$\cos\alpha - R_p\cos\alpha + R_s\sin\delta = 0 \quad \text{and} \quad \sin\alpha + R_p\sin\alpha + R_s\cos\delta = 0.$$

The reflection coefficients that satisfy these two equations are:

$$R_p = \frac{\cos(\alpha + \delta)}{\cos(\alpha - \delta)}, \quad \text{and} \quad R_s = -\frac{\sin(2\alpha)}{\cos(\alpha - \delta)}.$$

A similar approach can be used for an incident S-wave. Hence, even a rigid boundary leads to mode conversion.

(6) We need to calculate the following integral:

$$\sigma(t) = \int_{-\infty}^{t} g(t-\tau)\frac{\mathrm{d}\sigma^e(\varepsilon)}{\mathrm{d}\varepsilon}\frac{\mathrm{d}\varepsilon(\tau)}{\mathrm{d}\tau}\mathrm{d}\tau$$

$$= \int_{0}^{t}\left\{\sum_{i=1}^{N} h_i\,\mathrm{e}^{-\alpha_i(t-\tau)}\right\} AB\,\mathrm{e}^{B\gamma\tau}\gamma\,\mathrm{d}\tau.$$

Rearranging and taking the summation outside of the integral yields

$$\sigma(t) = AB\gamma\sum_{i=1}^{N} h_i\,\mathrm{e}^{-\alpha_i t}\int_{0}^{t}\mathrm{e}^{(\alpha_i+B\gamma)\tau}\,\mathrm{d}\tau.$$

On integrating and applying the limits, this leads to our final answer:

$$\sigma(t) = AB\gamma\sum_{i=1}^{N}\frac{h_i}{(\alpha_i+B\gamma)}\left(\mathrm{e}^{+B\gamma t}-\mathrm{e}^{-\alpha_i t}\right).$$

In the case that the strain remains at a constant value, $\varepsilon = \gamma t_0$ for $t > t_0$, then, for all $t > t_0$, the integral's limits are 0 to t_0 instead of 0 to t. This leads to the expression for the stress for $t > t_0$ being of the form

$$\sigma(t > t_0) = AB\gamma\sum_{i=1}^{N}\frac{h_i\,\mathrm{e}^{-\alpha_i t}}{(\alpha_i+B\gamma)}\left(\mathrm{e}^{(\alpha_i+B\gamma)t_0}-1\right).$$

Further reading

P. Howell, G. Kozyreff and J. Ockendon, *Applied Solid Mechanics.* Cambridge Texts in Applied Mathematics, Cambridge University Press (2009).

S.P. Timoshenko and J.N. Goodier, *Theory of Elasticity.* Third Edition, McGraw-Hill Higher Education (1970).

References

1. P. Ciarlet, *An Introduction to Differential Geometry with Applications to Elasticity*. Springer, 2010. ISBN 978-9048170852.
2. H. Ockendon and J. Ockendon, *Viscous Flow*. Cambridge Texts in Applied Mathematics, Cambridge University Press, 1995. ISBN 978-0521452441.
3. W. Sharpe, *Springer Handbook of Experimental Solid Mechanics*. Springer (2008). ISBN 978-0387268835.
4. R. K. Korhonen and S. Saarakkala, *Biomechanics and Modeling of Skeletal Soft Tissues, Theoretical Biomechanics, Dr Vaclav Klika (Ed.)*. Intech Open Access Publisher (2011). ISBN 978-9533078519.
5. Y. Fung, *Biomechanics: Mechanical Properties of Living Tissues*. Springer New York (2010). ISBN 978-1441931047.
6. S. Cowin and S. Doty, *Tissue Mechanics*. Springer, 2007. ISBN 978-0387368252.
7. H. Rich, M. Odlyha, U. Cheema, V. Mudera and L. Bozec, Effects of photochemical riboflavin-mediated crosslinks on the physical properties of collagen constructs and fibrils, *J. Mat. Sci.: Mat. Med.* **25**(1), 11–21 (2014).
8. J. Mark, B. Erman and M. Roland, *The Science and Technology of Rubber (4th edn.)*, Academic Press, 2013. ISBN 978-0123948328.
9. K. Aki and P. Richards, *Quantitative Seismology*. University Science Books (2002). ISBN 978-1891389634.
10. P. Howell, G. Kozyreff and J. Ockendon, *Applied Solid Mechanics*. Cambridge Texts in Applied Mathematics, Cambridge University Press, 2009. ISBN 978-0521854894.
11. K. Dussik and D. Fritch, Determination of sound attenuation and sound velocity in the structure constituting the joints, and of the ultrasonic field distribution within the joints on living tissues and anatomical preparations, both in normal and pathological conditions, *Public Health Service, National Institutes of Health Project.* **A454** (1956).
12. E. Madesen, J. Zagzebski, R. Banjavie and R. Jutila, Tissue mimicking materials for ultrasound phantoms, *Med. Phys.* **5**(5), 391–394 (1978).
13. P. Laugier and G. Haïat, Introduction to the physics of ultrasound. In *Bone quantitative ultrasound*, pp. 29–45. Springer, 2011.
14. M. W. Urban, S. Chen and M. Fatemi, A review of shearwave dispersion ultrasound vibrometry and its applications, *Current medical imaging reviews.* **8**(1), 27 (2012).
15. C. Verdier, Review article: Rheological properties of living materials: From cells to tissues, *Journal of Theoretical Medicine.* **5**(2), 67–91 (2003).
16. E. Detournay and A. Cheng. Fundamentals of poroelasticity, In *Comprehensive Rock Engineering, Vol 2: Analysis and Design Methods, (Eds. E.T. Brown, C.H. Fairhurst & E. Hoek)*, pp. 113–171. Pergamon Press, 1993.

17. A. Verruijt, *Theory and Problems of Poroelasticity*. Delft University of Technology (2013). URL http://geo.verruijt.net/.
18. B. Wirth, I. Sobey and A. Eisenträger. A note on the solution of a poroelastic problem. Tech. rep, Oxford University Mathematics Institute, Numerical Analysis Group, (2010). URL http://eprints.maths.ox.ac.uk/901/1/NA-10-01.pdf.

Printed in the United States
By Bookmasters